The Design Dimension

TO CLARE BURY
without whom this book would never have been written

The Design Dimension
The New Competitive Weapon
for Product Strategy and
Global Marketing

Christopher Lorenz

Basil Blackwell

First published 1986
Reprinted 1986
First published in paperback 1987
This edition, revised and updated, first published 1990

Basil Blackwell Ltd
108 Cowley Road, Oxford, OX4 1JF, UK

Basil Blackwell, Inc.
3 Cambridge Center
Cambridge, Massachusetts 02142, USA

British Library Cataloguing in Publication Data
A CIP catalogue record for this book is available from the British Library.

Library of Congress Cataloging in Publication Data

Lorenz. Christopher, 1946-
    The design dimension : the new competitive weapon for product strategy
    and global marketing
Christopher Lorenz. -- Rev. and updated.
        p.   cm.
    Includes bibliographical references.
    ISBN 0-631-17748-5
    1. Design, Industrial.   I. Title.
TS171.L67 1990
658.5'752--dc20   90-34918   CIP

Typeset by Advance Typesetting Ltd, Oxfordshire
Printed in Great Britain by T. J. Press Ltd, Padstow, Cornwall

Contents

Foreword

Tom Peters, co-author of *In Search of Excellence* and author of *Thriving on Chaos*

'The necessary search for new and lasting bases of competitive advantage has no better starting point than these pages.'

In 1960, Christopher Lorenz reminds us, Harvard marketing professor Ted Levitt declared that 'the entire corporation must be viewed as a customer-creating and a customer-satisfying organism'. Three decades later, Lorenz concludes that all too few firms have gotten Levitt's message. Attaining a true market orientation is almost as elusive as ever. The problem: the costs of failing to make the shift are going up rapidly.

We are confronted, says Lorenz, with a 'violent shift' toward globalization, unprecedented shrinkage in product life cycles and the fragmentation of all markets. As a result, the pursuit of fundamental bases for product differentiation becomes more frenzied with each passing week. Yet the odds of achieving such fundamental differentiation have become lower and lower. To mention but one reason, almost all are now plugged into the world of new technologies – leapfrogging with technology, or maintaining an advantage via technology (as did, say, Digital Equipment with its early minicomputers), is a virtual impossibility.

Though Lorenz does a thorough and commendable job of detailing these forces, they are merely the preliminary bout preceding his main event: laying out a new role for industrial design.

Lorenz contends, with a wealth of supporting evidence, that design is still an afterthought at most firms. Designers – seen as mere 'stylists' – are traditionally called upon late in the product development cycle, to round off the corners and cover up the big, awkward motors that the engineers have cobbled together. The crime is the lost opportunity. Design, if wholly reconceived, can be a principal basis for achieving lasting product differentiation. The crucial point is wholesale reconception.

On the one hand, paying attention to design early in the product development process is called for. Access to top management for the designer is also a plus. But there's much more. Drawing upon the experience of firms such as Sony, and more recently Ford, Lorenz persuasively argues that the design function can be no less than coach, catalyst, orchestrator and integrator – surfacing new product ideas on its own and becoming, at times, the principal coordinator for the entire product development process.

Borrowing from the Netherlands' Philips, Lorenz presents a model for product development in which the 'industrial design vision' (aesthetic knowledge, social and cultural backgrounds, ergonomic requirements) is an equal intellectual and organizational partner to the 'engineering vision' (technical research, production methods) and the 'marketing vision' (market research, market analyses, distribution systems). Precisely such equality, for instance, lies behind the success Ford has had with the Taurus and Sable.

The book begins with first principles: a stunning section, for example, systematically examines why so few firms have adopted a true market orientation (which is, in most respects, a prerequisite to a thoroughgoing design orientation). But the meat of this thin, readable volume is six superb case studies. On the one hand, Lorenz provides cases about firms that have lived – and benefited from – a design obsession almost from the start: Italy's Olivetti and US farm equipment-maker John Deere. The last four analyses are of those who have come more recently to an awareness of design's *strategic* potential: Sony, Ford, Philips and Baker Perkins. Baker who?

The latter, a mid-sized British producer of process machinery, was a most unlikely candidate to undergo, or benefit from, a design 'revolution'. But that's the point: in its less-than-style conscious market, design (broadly conceived) still 'works', and can add immense value – especially as Baker Perkins (now as part of the APV group) moves to become a true global marketer.

The most important part of the cases is the alternative path to design ascendancy which each represents. Sometimes outside designers are close to the chief executive (Olivetti). Sometimes design virtually runs the show, at least for part of the time (Sony). Sometimes a major reorganization gives design equal footing with the other main players at the strategy-making table (Ford). Fortunately, there is no set formula, no set path; any number of approaches can foreshadow success.

Lorenz concludes with an original (as original today as when the book was first published) and cautionary note about going overboard on global branding. The author is an effective, almost strident, advocate for the necessity of thinking globally – for firms of all shapes and sizes. But he does not fall into the common trap of translating this into a vision of a world with mindless, homogenized tastes, where one size fits all. Local adaptation, as often as not, will be more important than ever, he contends.

'The creation of real differentiation has become even more important than in the past', Lorenz concludes. 'But the old weapons for achieving this have become inadequate.'

The power of the book is its blend of strategic vision and telling detail. While the cases provide illuminating and practical evidence, the book's chief purpose is to mobilize management to consider a broad-scale, new conception of the firm in pursuing competitive advantage – via design – in the turbulent years ahead.

Introduction

'Design' means different things to different people. Depending on their point of view, it conjures up an image of women's fashions, designer clothing, furniture, fabrics and interior design, or even crafts. To some it embraces architecture, to others the creative side of engineering: design engineering. To very few does it suggest the activity which spans both the form and function of manufactured products – industrial design.

It was to fill this gap in public perceptions, and to explore the commercial potential of industrial design, that this book was conceived. I wanted in particular to examine the growing claim of industrial designers to be capable of far more than mere styling – 'the wrapping of products in nice shapes and pretty colours', as one cynic described their traditional role. Creating products whose shape, feel and appearance give satisfaction and pleasure to the consumer is certainly an element of the industrial designer's work; feel and appearance are part of the function of most manufactured products. But, whether the industrial designer has been trained at art college (as in America and Britain), or in architecture (as is often the case in continental Europe), is his or her contribution limited to the creation of form?

The idea that industrial designers might be just as capable as marketing experts and engineer–technologists of conceiving new products, and of playing a full and sometimes leading part in their successful development, provoked much initial scepticism in the breast of a hardened observer of the 'heavy' sides of management: corporate strategy, planning, finance, research, production and marketing.

Yet all the evidence supports the claim. Initially from winners of the British Royal Society of Arts Presidential Awards for Design Management (now run jointly by *The Financial Times* and the London Business School), and then from several famous international companies elsewhere in Europe, as well as the US and Japan, it emerged that industrial design was indeed beginning to be used in such ways. A handful of companies had done so for decades, but suddenly more were joining them.

The purpose of this book is to describe the emerging role of industrial designers in such companies, and to set it against the background of changes in the pattern of marketing and corporate strategy, especially the violent shift towards 'globalization'. Without an understanding of the way marketing has developed – or, more often, failed to develop – as a central driving force of corporate strategy, it is difficult to grasp the realities of industrial design.

Rather cursory treatment is given to the other elements of the 'design spectrum' within industrial companies: engineering design, environmental design (factories, offices, retail stores), and communications design (packaging, graphics, corporate identity). This is not because of any belief that these facets have a lesser commercial value. Nor does it imply that industrial design can be managed in isolation: just as the designer has to work in a team alongside marketing and engineering specialists, so he or she must have a particularly close relationship with packaging design and with the creators of the company's corporate identity. Not only is the importance of these elements more widely recognized than that of industrial design, but most of them have been well analysed by other authors.[1-3] The same applies to all the moral questions which surround the social value of design; they have been debated ever since Victor Papanek's swingeing attack in 1971 on American industrial designers for having 'elected to serve as pimp for big-business interests'.[4]

Though obviously concerned in the first instance with the experience of companies which design and make products, the findings of the book also reach deep into the service sector: differentiation by design is, or ought to be, of central concern to the worlds of banking, financial services and retailing, as well as to any marketing or advertising department. There is much for others to learn from the successes and failures of design management in industry.

Chapter 1 sets the scene, outlining the issues covered in subsequent pages, and presenting initial evidence for the claim that industrial design really can be much more than styling. Its emerging role as a key competitive weapon is examined in the light of the new priority which companies are having to give to marketing, and to the development of products for global markets.

The second chapter clarifies the historical meaning of the term industrial design, traces the development of the design profession to the present day, and analyses the demanding nature of the design process. In examining the factors behind the narrow image in which industrial design has traditionally been held, it is suggested that this is partly due to the way the industrial design profession developed in the 1920s and 1930s, under contrasting transatlantic influences: elitist Bauhaus austerity in most of Europe; Madison Avenue and Detroit excess in America. The inferior status which these beginnings conferred upon industrial designers (with only a few notable exceptions) was reinforced, rather than reversed, by the development of marketing and market research. Yet the visualizing and synthesizing skills of designers were increasingly in evidence for those who wished to see them.

Chapter 3 discusses the all-important marketing side of the equation: the puzzling reasons why it has taken over a quarter of a century for marketing really to take root in most Western companies. The explanation lies partly in the very complexity of what is known as the 'marketing concept', partly in the way marketing has often been confused with its supporting techniques (especially market research) and partly with the many other pitfalls into which companies can – and do – fall.

Chapter 4 examines the factors which since the mid-1970s have combined to stimulate a widespread conversion to *real* marketing. The very same set of stimuli has also created the need for a radical upgrading in the role of industrial design – not only to supplement the elusive 'imagination' of the marketing function, but also to assist in the speeding-up of product development in order to meet the challenge of global competition. One of the key weapons in this battle is better design-for-manufacture, so that costly and time-wasting design changes do not have to be made halfway through the development process, or after production starts.

Part II (chapters 5 – 11, which have been heavily updated for this revised edition) contains a range of contrasting international case studies of 'design unchained' – various ways in which industrial designers are assuming a much enhanced role in major companies, either as part of a commitment right across the 'design spectrum', or as a more focused concern with product design. Two pioneering design-minded companies are examined in depth: Italy's Olivetti and America's Deere & Co (best-known for its smart green and yellow John Deere tractors and harvesters, but also noteworthy for its belief in excellent environmental design). Design management within four recent converts is then put under the microscope: Sony in Japan; Ford (in both the US and Europe); Philips (likewise); and, the joker in the pack, the remarkable Baker Perkins of Peterborough, England, which from the early 1970s found industrial design of enormous value in helping it achieve prominence in, of all things, printing and baking machinery. Since 1987 it has been part of the APV process engineering group, which acquired it partly for its strengths in the management of design and technology.

The section concludes with a look at the designer at work, through the unrivalled experience of Kenneth Grange, one of the founder-partners of the world-renowned Pentagram consultancy. Grange has worked for a wide range of companies, from Kodak to a Japanese sewing machine maker, and he has held several influential posts as 'consultant design director'.

Part III – also updated for this edition – sets these developments in the context of the much-hyped (but still little understood) trend towards 'globalization' of products and markets. Chapter 12 clarifies the war of words over globalization by exploring its promise and its limits, as applied to both products and brands. (Surprisingly enough, many of the world's leading companies and advertising agencies seem to have forgotten that the two are not identical – this is a very dangerous error indeed.)

Chapter 13 sums up with an assessment of whether globalization and other current trends in technology, marketing and design itself (including computer-aided design) create an opportunity for further enhancement of the role of design, or pose a threat to the designer's influence. The answer is evenly balanced: as with the emergence of design in the pre-global era, it depends both on the designer's ability to develop new skills, and on top management's readiness to recognize them. Much also rests on the degree of care with which the challenge of globalization is handled by management.

The book concludes with an appendix which examines one of the 'secrets of success' of Japanese industry: the strategy of 'cascading' from one related segment to another, until a whole market has been annexed.

I am most grateful to those whose actions and evangelism sparked off the writing of the book. They include: the late Lord Wilfred Brown, a company chief executive and government minister who rampaged around British industry in the 1960s and early 1970s, almost as a lone voice, demanding 'What are *you* doing to promote design?'; Sir Terence Conran, the swashbuckling designer–retailer; and James Pilditch, founder of Aidcom International and the doyen of British design consultancy. In different ways, all three could claim to have laid the ground for the British design revival of the 1980s. In a country whose design illiteracy is virtually without parallel, except in the many art and design colleges which have somehow managed to thrive in a sea of social scepticism, such a phenomenon is remarkable indeed.

Particular thanks are due to Sir Geoffrey Owen, Editor of *The Financial Times*, for encouraging me to write about design long before it became fashionable to do so; also to Nicholas Leslie, Kenneth Gooding and other colleagues on the 'FT' for their support.

My debt to Professors Theodore Levitt and Philip Kotler, respectively of the Harvard Business School and Northwestern University, is evident from the text of the book, as is the invaluable assistance I received from the companies and designers whose work is described in part II. Among the many individuals working for those companies who are unsung elsewhere in these pages, but who made a major contribution to them in terms of patient and expert advice, are David Maroni of Olivetti, Mitsuru Inaba and Akihiko Amanuma of Sony, Peter Nagelkerke, Frans van der Put, Mike Jankowski and Margaret van Alberda of Philips, and John McConnell of Pentagram.

Part I
Making New Connections

1 The Power of Design

Even by the extraordinary standards of Hollywood, it was one of the most lavish parties ever thrown. To the tune of 'Happy Days are Here Again', more than 1000 guests from the worlds of movies, TV and high society were wined and dined amid rampant luxury in the historic MGM studio where *Gone With the Wind* had been filmed.

The occasion of this $1.5 million extravaganza was not a film premiere, but the 'roll-out' of the 1986 Ford Taurus and Mercury Sable, two sleek new cars on which the Ford Motor Company had pinned its hopes for a competitive revival against the massed ranks of General Motors, the Germans and the Japanese.

As the world now knows, the Taurus and Sable did indeed succeed in propelling Ford into a position of unexpected prosperity in the late 1980s, putting it in a much better position than GM to fight the tough competitive battles of the 1990s.

The Taurus/Sable launch also had a wider significance. It marked the conversion of the world's second largest motor company to a strategy of competing through adventurous, aerodynamic product design. Gone was the traditional policy, common to all American motor manufacturers, of cladding a lacklustre and unimaginative vehicle in an unwieldy, boxy, battering-ram shape, garnished with all sorts of ritzy, angular radiator grilles, tail fins and chromium strips. In its place was a policy of integral design, in which the car's uncluttered shape was heavily influenced by its function, and particularly by the need to reduce wind drag in order to improve its fuel consumption.

The strategy had been sparked off by Ford's European offshoots in the late 1970s. Conservatives back home in Dearborn, Ford's headquarters in the heart of mid-west Michigan, had taken time to become convinced by it. But, stung by the company's poor sales and its plunge into heavy financial losses between 1980 and 1983, its top brass were now committed. As Donald Petersen, Ford's new chairman, told the gathering of celebrities, the company's 'dynamic vehicle philosophy' (the motor industry just loves hyperbole) encompassed not only the virtues of performance, handling and aesthetics, but also a galaxy of characteristics like quality, function, safety, comfort, reliability and cost of ownership. (The detailed story of Ford's successful design-led revival is told in chapter 8.)

In shifting to this unusually deep commitment to product design, Ford had to undergo a conversion of Galileo-like proportions. Conventional wisdom in the automobile industry had always put a company's interests before those of its customers. 'But that has changed' declared Petersen. 'Now the driver and passengers, not the company, are the center of Ford's universe.' If the result was a set of products that made Ford markedly different from its competition, that would no longer be a worry – in fact, so much the better. Never mind if their low noses, high tails and smooth shapes quickly earned them the nickname of the 'jelly-bean look'.

In down-to-earth terms, Petersen was saying that Ford's use of design as a competitive weapon formed part of its belated conversion to a broader cause: the concept of marketing first popularized more than a quarter of a century ago by Theodore Levitt and others, in which the imaginative satisfaction of consumer needs and wants, whether active or latent, takes over as the company's driving force from the traditional approach of trying to sell whatever the company happens to produce. Remarkably, this shift from 'sales' to 'marketing' is one which many companies have been slow to make. They may have given their sales chief a grand new 'marketing' title, but they have continued to lack the ability to think long term, to think not in terms of an amorphous mass market but of particular market segments, and to be imaginative in the identification of potential new segments and products.

Only in the early 1980s did things really start to change. From Tokyo to Detroit, Milan to Munich, London to Los Angeles,

From boxy, unwieldy styling to sleek functional aerodynamics: Ford's leap forward into integral design. Above: the traditional American look, early 1980s (Ford Crown Victoria). Overleaf: the new 'Aero' generation, 1986.

companies large and small belatedly began to embrace 'the new era of marketing', in which product design is used as a key competitive weapon. They were recognizing that, for those involved in competition on a global scale – a rapidly increasing proportion of companies – the design dimension was becoming a particularly important factor. Hence the way that, as the 1980s neared their end, it was being exploited more and more to create competitive distinctiveness for products of all kinds, whether they were Minolta cameras or Sony hi-fis from Japan; Philips compact disc players or shavers from Holland; Wilkinson razors from Britain; Audi automobiles from Germany; or the 'Swatch' watch from Switzerland.

A familiar strategy in premium products, for example Rolex watches, Braun shavers, Porsche cars and Herman Miller office furniture, this form of differentiation has been spreading like wildfire to the world of mass marketing in the last few years. In the words of Levitt's fellow guru, Philip Kotler, 'one of the few hopes companies have to "stand out from the crowd" is to produce superiorly designed products for their target markets'.[5]

Design is no longer a luxury, in other words, but a necessity.

Targeted at the booming 'Yuppy' market: the Mercury Sable.

To some of the latest design converts this is merely a matter of styling: the first $30 'Swatch' was essentially a stripped-down and zappily restyled version of the much more expensive Concord Delirium, which was the thinnest watch in the world when it was launched in 1979. The new cars which in 1982 and 1983 heralded Ford's initial shift to aerodynamics in the US – the Thunderbird, Lincoln Continental Mark VII, Tempo and Topaz – and which boosted its market share dramatically, were also reskinned versions of existing models.

But, like Ford's European range of cars – especially the Sierra and Scorpio/Granada – its post-1985 US models reflect a radical redesign of what lies under the skin. As Kotler argues, in order to succeed, a company must 'seek to creatively blend the major elements of the design mix, namely performance, quality, durability, appearance and cost'.[6] Each of the elements affects the others, and it is becoming unacceptably expensive, in competitive as well as financial terms, to decide them separately – they have to be specified in parallel, with all the necessary trade-offs settled at the start. Yet many companies persist with the conventional pattern of leaving decisions about the various aspects of industrial design until last.

It is through their recognition of the need to manage product development along Kotler's lines that the more enlightened

The Sable's stable-mate: the 1986 Ford Taurus station wagon.

companies have begun to realize they must stop treating industrial design as an afterthought, and cease organizing it as a low-level creature of marketing (whether marketing 'proper', or sales masquerading under another name). Instead, having overcome their long-standing doubts about the 'seriousness' of art- and architecture-based designers (as opposed to engineers), they have elevated industrial design to fully-fledged membership of the corporate hierarchy, as it has been for decades in design-minded companies such as Olivetti, Deere & Co and IBM. (Paradoxically, these three all rely at least in part on outside design consultants, rather than just on in-house teams. But the consultants are so well integrated by now that they are treated as insiders – with the difference that their external experience wins them extra respect.)

Other companies have gone even further by recognizing that design is so central to the company's purpose, and such a multi-disciplinary skill, that industrial designers can play a catalytic role in the product development process, and even, through product strategy, in helping form market strategy. Again, this

applies not only to Braun, Porsche and other 'minority' manufacturers. Sony, Olivetti and even the much larger Philips can all
boast successful products which were conceived by industrial
designers working informally as product planners and project
leaders.

Usually this happens behind the scenes, but in some cases they
have taken on this role officially. During the late 1980s, for
instance, Sony's design chief was given the additional role of
coordinating the development of products which combined the
expertise of the company's various organizational groupings, such
as its audio, video and television divisions. In the electronics
industry this sort of integration role has become more and more
crucial with the growing popularity of home video 'systems'
which combine innovative audio techniques (such as the compact
disc) with top-class video, and with interactive home computer
systems.

Even in the most unlikely of industries, industrial designers are
helping to coordinate the product development process. As long
ago as 1981 the chief executive of Baker Perkins, a leading international process machinery maker, was describing his industrial
designers as 'translators, bridges and catalysts' between marketing
and the various types of engineer: design engineer, development
engineer, production engineer and so on. In common with top
managers in a growing number of other companies since then,
he was using his industrial designers as an extra arm of general
management.

In the arcane but descriptive language of behavioural science,
industrial designers who are given such pivotal roles display a
combination of several skills which are generally considered
necessary to the success of any management team. At one and the
same time they seem to be acting, alone or in conjunction with the
official project team leader, not only as an invaluable source of
ideas, but as 'facilitator', coordinator, evaluator and completer.
This is a very far cry from the stereotype of 'designer as stylist', and
much closer to the all-round role of coordination and integration
which, in many countries, an architect plays in the building
process.

That sort of project management is just as important in the
process of designing, developing, making and launching a
manufactured product, but it is seldom managed successfully.

Coordination of all the different specialists is either left to a very formalized but inefficient procedure of interdepartmental communication, or is concentrated in the hands of a project team leader, programme manager or product manager who frequently lacks imagination and has an inadequate understanding of the various specialist skills at his or her command. In the words of one senior manager within the sprawling Philips organization, 'good product managers are a very rare breed'.

This dearth is underlined by a six-year Harvard Business School study into product development in the world automobile industry; one of the many conclusions of the research, which was completed in 1990, was that a prime difference between average and best product development performers in the Japanese auto industry was that product managers in the most effective companies acted, in the researchers' words, 'like designer–integrators'.[7]

Underpinning the ability of many industrial designers to play a full part in the development team, and the potential of some even to become the team's coordinator – official or otherwise – is a set of unusual personal attributes and skills. Some are inborn, others are learned. They include imagination; the ability to visualize shapes and the relationship between objects, in three dimensions; creativity; a natural unwillingness to accept obvious solutions; the ability to communicate, through words as well as sketches; and, finally, the designer's stock-in-trade – the ability and versatility to synthesize all sorts of multi-disciplinary factors and influences into a coherent whole.

This is a pretty demanding combination – even more so when the successful designer in industry must also possess all the usual executive virtues of determination, drive and discipline. But many architects possess it: not those notorious individualists with massive artistic egos, but the more level-headed, cooperative and business-like variety.

So do a good number of industrial designers, regardless of whether their training was at architectural or art school. Contrary to popular myth, these latter-day versions of Renaissance Man (and Woman) do not all hail from Milan and the other design-rich parts of Italy. Ettore Sotsass, Mario Bellini, Rodolfo Bonetto, Giorgetto Giugiaro and other famous Italian consultants are undoubtedly international masters of design, but so are Germans such as Richard Sapper and Hartmut Esslinger, as (respectively) IBM and both Apple and Sony have recognized in the form of lucrative contracts. A number of less well-known Britons, too, deserve the same accolade: not only consultants such as Kenneth Grange and Nick Butler, who have worked, respectively, for Kodak and Minolta cameras, among many other clients, but also a bevy of unsung heroes who operate as insiders at world-scale companies such as Ford, BMW and Olivetti. British and German designers seem to take more easily than their Italian counterparts to the culture and discipline of working within a large company. So do Americans – which is just as well, since many of the top US product designers are anonymous members of in-house teams. With a few exceptions, such as Niels Diffrient, a former partner of the late Henry Dreyfuss, most of the great American product design consultants are names from the past.

Whatever their nationalities, few other sorts of professional can be expected to possess such a broad combination of characteristics and skills, be they planners or accountants, engineers or even marketing executives. Indeed, it is the very paucity of vision on the part of many so-called 'marketing' departments which often torpedoes the application of Levitt's new formula, 'the marketing imagination'.

The clear message of this book is that, for a company to develop a fully-fledged 'marketing imagination', and to exploit it to its utmost, it needs to upgrade its use of design. As Levitt himself argues, 'the search for *meaningful distinction* [my italics] is a central part of the marketing effort'. Yet, in a crowded and increasingly global marketplace, the achievement of meaningful distinction requires the company to make all sorts of new connections.[8]

In the broadest of senses, it must make new connections between itself and the consumer. To do this it must be able to establish more effective links within its own organization between the various elements of the company's 'value chain' (or 'business system', as it is sometimes called), notably: technology which is either available on the market or is coming out of research; development; production; marketing; sales and distribution; and service. Of particular relevance to the subject of this book, it must make new connections between the market and the various elements of Kotler's 'design mix': performance, quality, durability, appearance and cost. Again, the Harvard automobile study supports the importance of making such internal and external connections (though the authors prefer the term 'internal and external integrity').[9]

For these connections to be made successfully requires a team effort in which the industrial designer's imagination, synthesizing skills and entrepreneurial drive are given as much weight as the tools of the engineer, the financial controller and the marketer. The design dimension is no longer an optional part of marketing and corporate strategy, but should be at their very core.

2 More than Just a Pretty Face:
The Roots and Range of Industrial Design

Industrial design is a decidedly twentieth-century phenomenon. Its roots lie in the separation between design and production which occurred in the Industrial Revolution – until then the craftsman who designed an object usually also made it, either himself or with the help of his workshop assistants.

But it was not until after 1900 that the industrial design function really began to develop. Confusingly, the term does not apply to the entire process of design in industry. It was originally coined as a grander substitute for the phrase 'industrial art'. The more technical aspects of design, which in the nineteenth century were frequently called 'invention', took on the term 'engineering design' or 'design engineering', though the more mundane activities within the design process are often described as 'technical drawing'.

From their respective sides of the fence, today's engineers and industrial designers often lapse into use of the simple word 'design', as do general managers and marketing executives – not to speak of members of the general public. But they are normally referring to a quite specific part of the 'design spectrum': engineering design, industrial design, graphic design, textile design or one of the spectrum's many other elements.

The earliest recorded official use of 'industrial design', with its specific meaning, came in 1913, when the US Commissioner of Patents proposed a modification of regulations to protect property in industrial design. The phrase was used quite clearly as a generic description for the distinguishing *form* of products, as distinct from their *function*.[10]

Yet the birth of industrial design is often dated six years earlier, and located in Germany rather than the US. In 1907 several German companies had commissioned a number of craftsmen and architects to design various products for machine manufacture. Among the commissions was one from AEG, the large electrical company, to Peter Behrens – who went on to work for many years on almost every aspect of the company's design: not only its products, but also its graphics and buildings.

From the very start, a yawning gulf developed between European and US conceptions of industrial design: the one highly intellectual and dedicated to functional simplicity (what has often been described as 'working from the inside out'), the other a styling tool at the service of sales and advertising, where the exterior was all-important, and the inside mattered little.

Behrens had all the right European credentials. He was closely involved with the Deutscher Werkbund, an organization with high theoretical ideals which was set up in 1907 by the Prussian government to develop a 'new aesthetic' for machines. Influenced by the Arts and Crafts Movement of late-nineteenth-century Britain, the Werkbund set out on a quest to restore dignity to work in the new machine age. One of its main tenets was that there was such a thing as an absolute standard of 'good design'. In 1915 an official body with similar objectives, the Design and Industries Association, was founded in Britain. Such worthy bodies have always been noticeably absent in the US.

Apart from Behrens' work at AEG, neither achieved much direct effect in industry. But the Werkbund nevertheless left an immeasurable legacy for European design. In 1919 it spawned an organization whose impact reverberates to this day: the Staatliches Bauhaus. Led by Walter Gropius and later by Ludwig Mies van der Rohe, both of whom had trained in Behrens' office (along with the influential French architect Le Corbusier), the Bauhaus developed a set of challenging theories that went beyond mere functionalism. Emphasizing the importance of geometry, precision, simplicity and economy, it provided the intellectual underpinning for more than half a century of architectural education and practice under the banner of the 'Modern Movement'.

Via AEG, Olivetti and Braun, the doctrine of truth to materials and functional design found its way into the products of a broad swathe of European companies. But it was a slow process. Initially

the elitist nature of Bauhaus doctrine deterred the attentions of hard-nosed industrialists. Only a few intellectuals with companies at their command bothered to turn the theories to practical use. The most influential was Adriano Olivetti, an ardent student of architecture who also put Bauhaus ideology into practice through ambitious urban planning schemes.

It was not until the late 1950s, amid the post-war recovery of the European economies and the growth in business competition, that industrial design really began to figure in the priorities of most managements. Then, using either internal design staffs or the growing band of design consultants, they quickly adopted one of several variations of the doctrine of functionalism. Some went for fully-fledged German austerity, following the example of the Braun brothers who had fallen under the Bauhaus-like spell of the Hochschule für Gestaltung at Ulm, and had employed one of its star graduates, Dieter Rams. Others took the slightly softer Scandinavian line, in company with Danish furniture, Electrolux vacuum cleaners and Saab cars.[11] Still others tried to emulate the slightly more 'arty' Italian approach. The motor industry was inevitably influenced by American styling trends, but even it paid considerable attention to functionalism.

In some cases, this functionalism was little more than styling – but it was styling that bore some relation to the product's purpose. And in quite a number of companies it was very much more than styling. At first sight a streamlined European motor car or locomotive of the 1930s might look similar to its American counterpart. Yet the differences were considerable. European shapes were generally much cleaner and smoother, and heavily influenced by scientific theories of streamlining for greater speed and increased stability. Whereas the American use of streamlining tended towards excess, such as the redundant use of 'aerodynamic' aircraft motifs on automobiles, and even streamlined coffins, many European industrial designers worked hand-in-glove with engineering designers to create a powerful synthesis of aesthetics and technology.[12]

Left: *Contrasts in streamlining: clean functionalism (top right and centre right) versus skin-deep protruberance and cynical excess (left and bottom). Top right: A4 Pacific steam locomotive UK 1935. Designer: Sir Nigel Gresley. Centre right: Chrysler 'Airflow' car US 1934. Designer: Carl Breer. Left: Pennsylvania Railroad Engine 3768 US 1936. Designer: Raymond Loewy. Bottom: General Motors Cadillac 'Eldorado' Brougham US 1955. Designer: Harley Earl. Drawing by Michael Daley.*

European functionalism did make an early mark in the US, but only in design education and architecture. By the time Gropius, Mies and several other Bauhaus luminaries had taken up US residence during the 1930s, having been driven out of Germany by Nazism, the American industrial design profession was fully established and well on its way to commercial influence – but in precisely the opposite direction to functionalism.

Whereas the first European industrial designers were architects and engineers, most of America's pathfinders were neither. Men such as Walter Dorwin Teague and Norman Bel Geddes were theatre designers and artist–illustrators who seized on the growth of industrial competition in the 1920s, and in the Depression of the early 1930s, to offer their services as an adjunct to advertising; in the words of one design historian, they 'extended advertising into the product itself'.[13] Even Raymond Loewy, a French engineer who had emigrated to the US, began work as a fashion illustrator and theatre designer.

Looking back on the beginnings of American industrial design from the still-fresh vantage point of 1934, *Fortune* magazine reported that 'Furniture and textiles, their usefulness taken for granted, had long sold on design. Now it was the turn of washing machines, furnaces, switchboards, and locomotives'.[14]

One of the first to get the message was General Electric, which in the early 1920s established a committee for 'product styling'. As American design historian Arthur J Pulos reports, many manufacturers quickly realized that the appearance of the product in an advertisement would be an important element in its public acceptability, and this placed the advertising agency and its artists in the position of having to make certain that the product being promoted was as attractive as it was useful.[15]

Despite the resistance of Henry Ford, this message was not slow to permeate the automobile industry. In 1926 Alfred Sloan at General Motors stole a march on Ford by discovering the selling power of making annual changes to the shape and decoration of his products. Ford was forced to swing into line, and the die was cast: Detroit was wedded from that moment to a strategy of planned obsolescence through skin-deep design – an alliance which German critics have called 'Detroit Machiavellismus'.

The introduction of this approach in the automobile industry and elsewhere was accompanied by the rapid development of market testing, in which consumers were asked to say what they thought of different products, both in general and in detail. From this very basic form of market research, which was pioneered as part of his design activities by, among others, Norman Bel Geddes, there developed more than half a century of market research myopia: manufacturers slavishly followed the results of market tests which, by their very nature, could merely reflect the consumer preferences of today – not the possible patterns of tomorrow.

So advertising and the appeal of theatrical make-believe ('the creation of dreams', in the advertising industry's parlance) were the prime stimuli behind the establishment of industrial design in the US.

This pattern was reinforced in the mid-1930s by the commercial failure of an automobile which, for once, reflected a 'European approach to design': the Chrysler Airflow. Design writer and historian Stephen Bayley judges it to be one of the first attempts which American industry had ever made to coordinate engineering and appearance in a single package. It was the last for a long time. The industry went back to skin-deep styling – with a vengeance.[16]

The arch exponent of the Detroit approach was Harley Earl, GM's director of styling and a showman par excellence, whose thinking dominated the company's visual policies from his arrival in 1927 until well after his death in 1954. His most famous creation was the tail fin, a motif which he borrowed from the Lockheed Lightning fighter, but which improved the performance of his automobiles not one bit – if anything, the reverse.[17]

The fast-growing fraternity of US industrial design consultants adopted a basically similar approach to Earl's. But an exceptional few were influenced by European ideas. Prime among them was Henry Dreyfuss, who was uniquely concerned with the fitting of products to people. As such, he is still seen by many as having been 'the conscience of the design profession', not just in its early years, but right through to his death in 1972.[18]

Dreyfuss complained from the start about Detroit's habit of disguising what he called 'good form' (form that was related to function), 'with an overabundance of chrome teeth, disks, wings and meaningless shiny bands'.[19] He was equally critical of the 1930s fashion for streamlining – though he himself had helped create it. Looking back from the early 1950s he attacked the 'ridiculous' streamlining of fountain pens, baby buggies and pencil sharpeners. He preferred the concept of 'cleanlining', in which a streamlined appearance was directly related to function. He was appreciative, for example, of the way the design of toasters had been improved, through the junking of 'useless protruberances and ugly corners that not only spoiled good honest lines but interfered with efficient operation. Stand a 1929 toaster, with its knobs and knuckle-skinning corners and impossible-to-clean slits and overall ugliness, next to today's model, and the difference is apparent'.[20]

When he was asked in the late 1920s by both Macy's department store and Bell Telephone to advise on how to provide products with forms, shapes and colours which would improve their sales, Dreyfuss repeatedly insisted that exteriors on their own were irrelevant, and that he would only work 'from the inside-out'. Upon this approach, closely related to the principle that form should follow function, Dreyfuss based decades of successful collaboration with Bell (subsequently AT & T), as well as with Sears, Hyster lift trucks and countless other large American corporations (his firm's work for Deere and Co is examined in chapter 6). Among his creations for Bell were several classic telephone handsets which are still in use in thousands of homes and offices across the world.

Dreyfuss was thus paying extraordinarily careful attention to the impact of products on the user well before the emergence of ergonomics as a discipline after the Second World War; indeed, he is himself widely credited with having helped establish the discipline with his books *Designing for People* (1955) and *The Measure of Man* (1961). In all his writing and his firm's work, anthropometric drawings of 'Joe and Josephine' figure prominently. They show every possible variation in dimension of an averagely sized American couple: trunk, legs, arms, vision distance and so on.

Describing Joe and Josephine as the hero and heroine of his 1955 book, Dreyfuss made crystal clear the depth of his belief that industrial designers should be concerned not only with appearance, but also with function. He quoted his firm's favourite maxim, that:

What we are working on is going to be ridden in, sat upon, looked at, talked into, activated, operated, or in some way used by people individually or en masse. If the point of contact between the product and the

World classics: three generations of Bell telephone handsets (US) designed by Henry Dreyfuss Associates. Top: '300' set, 1937; centre: '500' set, 1951; bottom: 'Trimline', 1965.

people becomes a point of friction, then the industrial designer has failed. If, on the other hand, people are made safer, more comfortable, more eager to purchase, more efficient – or just plain happier – the designer has succeeded.

This unusually thorough approach to design helps explain why Henry Dreyfuss Associates has been one of the few original US design consultancies to survive the death of its founder and the growing emphasis among client corporations on the use of in-house design teams.[21]

Equally influential, in the lead given to American companies which now wish to break out of the narrow confines of 'design as gaudy styling', was the work of Eliot Noyes for Westinghouse, Mobil and especially IBM.

Along with Charles Eames, well known for his Herman Miller chairs, Noyes was a breed apart from the traditional American industrial design community. Trained as an architect, and as such strongly influenced by Bauhaus thinking, he shared the Modern Movement's belief in simple, 'clean' design.

Noyes was hired in 1956 as Consultant Design Director by IBM, whose head, Tom Watson Jr, was a great admirer of the products of Olivetti, one of IBM's main competitors. Under Noyes, IBM's design was transformed into a model of classicism and restraint – and of systematic, corporate-wide management. Reaching right across the spectrum, from products to communications and environments, it made a considerable contribution to 'Big Blue's' real-life image as a solid, reliable, slightly up-market company. Since Noyes' death in 1977, his design policies have been more or less faithfully continued.

Given the success of Noyes' approach at IBM and elsewhere, it may seem surprising that it took until the 1980s for American industry to really begin to discover the attractions of linking form to function wherever possible, and (even in electronic products where form and function now bear virtually no relation to each other) of exercising visual restraint. A large part of the explanation must be that, even in the face of a surge in imports of functional designs from Europe and Japan, deep-rooted habits die hard, especially among conservative corporate executives.

A more positive factor behind the belated US shift to a more integral approach is that European design has recently become less austere and straitlaced – and therefore more attractive to Americans. Since the early 1970s the starkness epitomized by Braun has lost something of its influence in Europe, and there has been more of an attempt to appeal to the sensual side of the consumer's psyche: whether in colour or shape, Olivetti typewriters can no longer be said to look purely functional. Nor can German household appliances or cars.

Paradoxically, this loosening of the stays of functionalism has come at a time when European industrial designers have begun to become more involved in the functional side of product design and development. Though they have always tended to be more involved than their American cousins, most of them have nevertheless been treated until recently as servants to engineering, sales and market research. It is no coincidence that the German word for 'industrial design' is 'the creation of form' (*Formgebung*), and that until 1972 the foremost award for design in Britain was called 'The Duke of Edinburgh's Award for Elegant Design'.

In such a climate, only a few European industrial designers, whether consultants or members of an in-house team, were initially treated by their clients or employers with the respect they deserved. In addition to the 'super-stars', notably Dieter Rams at Braun and first Marcello Nizzoli and then Ettore Sotsass and Mario Bellini at Olivetti, it was just a handful of consultants, such as Kenneth Grange (see chapter 11) who were occasionally able to influence the entire development process.

One of Grange's successes in the 1960s was the design of a small kitchen mixer for the British Kenwood company – he still delights in the story of how his first design of blender sent its contents shooting up to the ceiling, because of the shape of the goblet. As with several of his more recent clients, especially Wilkinson Sword, he was fully involved in work on marketing, tooling and labour costs, as well as in decisions on production techniques. At Wilkinson he quickly became a catalyst for all sorts of marketing and product decisions.

Significantly, almost all the early 'design-aware' companies were managed by a paternalistic regime where the chairman or chief executive – who was often also the owner – took a strong personal interest in design. In a broader sense this lesson still applies; at Ford, Wilkinson and the other multinationals which have converted to the use of design as a major strategic weapon since the late 1970s, top management commitment has played a vital part.

But in the bad old days of the 1950s, 1960s and early 1970s, the designer's ability to inch his or her way forward in professionally managed and publicly owned companies depended heavily on the personal credibility that he or she could establish through long and strenuous efforts at the coal face. This was true whether the designer was part of an in-house team or an outside consultant.

In a few parts of Philips, for instance, individual in-house designers gradually established such influence with middle management that they were playing a key role in product development well before top management decided to upgrade the entire industrial design function in 1980. The same applies to the in-house designers at Ford of Europe (but not Ford US), to Baker Perkins and – in Japan – to Sony (see part II).

For design consultants, working from the outside, it was only when they managed to form unusually lengthy and fruitful associations with a client that they were – just sometimes – able fully to assert their remarkable combination of skills.

To repeat the *caveat* made in chapter 1, it is not every industrial designer who is capable of playing a central role alongside engineering and marketing in the product development process. Even fewer have the commercial acumen and experience to make a contribution to marketing strategy. But some do – thereby adding immeasurable value to the companies for which they work.

Kenneth Grange, who is a living antidote to the preconception that designers are arrogant individualists, makes the point in typically down-to-earth fashion. The particular skill of the industrial designer, he says,

is to reconcile the benefits of long associations [of learning and experience] with the cross pollination of ideas from one field to another. It is a difficult skill embracing problems of confidentiality, a wide knowledge of many engineering and fabricating skills, and awareness of consumer and cultural needs (in the widest sense), plus sympathy with financial evaluation.[22]

Second only to the designer's most fundamental skill – this synthesis (or 'cross pollination') from one field to another – are the closely associated abilities of imagination and visualization.

Theodore Levitt provides an admirably clear definition of imagination: the construction of mental pictures 'of what is or is not actually present, what has never been actually experienced'. To exercise the imagination, as he says, requires 'intellectual or artistic inventiveness' (usually both). Levitt is right when he says that the exercising of imagination in business must involve not only the shedding of constraints such as convention and conviction, but also the discipline to combine disparate facts or ideas into new amalgamations of meanings.[23] He may also be right that a proportion of marketing executives is capable of exercising and channelling that imagination. So, for that matter, may some engineers.

But Levitt is on highly controversial ground when he claims that 'anybody can do it'. The odds are heavily stacked against much imagination slipping through the nets of formality and procedure that characterize the way most people think, and the way most organizations behave. By the time the majority of people reach adulthood, their imagination has usually been swamped by the imposition of conventional frameworks of thinking; this process is accelerated by the stifling bureaucracy of most companies.

As in marketing and engineering, there are, of course, exceptions to this rule in almost every walk of life. But in the industrial design profession, as in architecture, the situation is reversed: imagination is the rule, and conventional thinking the exception. At school it is frequently the budding designer's originality and irreverence which dictates his or her rejection of economics, science and the other largely 'academic' subjects. The well-rounded design course then takes this creative talent and, through a carefully balanced educational programme, develops it. But the best courses also apply the practical disciplines that the designer will need in industrial employment.

Nor is this the full story. Imagination of the sort described by Levitt may not be the designer's monopoly. But, if one includes architects within the definition of 'designer', the crucial companion skill of visualization may well be. The 'mental pictures' in Levitt's definition of imagination are not necessarily visual; like those of many poets, they may be purely verbal, or abstract. In people with the skill to visualize, they conjure up pictures, in living shape and technicolor. But this is not so in everyone.

One of the most illuminating accounts of the power of visualization was given as long ago as 1883, in *Inquiries into Human*

Faculty, written by a cousin of Charles Darwin, Francis Galton. A visual image, the Victorian reader was told:

is of importance in every handicraft and profession where design is required. The best workmen are those who visualize the whole of what they propose to do, before they take a tool in their hands . . . Strategists, artists of all denominations . . . and in short all who do not follow routine, have need of it.[24]

Galton argued that, just like the village smith and carpenter, engineers needed this skill. The great designer–engineers, men like George Stephenson the locomotive builder and Isambard Kingdom Brunel the civil engineer, did indeed possess it. But Galton was surprised and disappointed to find that most 'men of science' claimed that mental imagery in the literal sense was unknown to them. He concluded that:

Our bookish and wordy education tends to repress this valuable gift of nature. A faculty that is of importance in all technical and artistic occupations, that gives accuracy to our perceptions, and justness to our generalizations, is starved by lazy disuse, instead of being cultivated judiciously in such a way as will on the whole bring the best return.

One hundred years later, the education system in most countries remains 'bookish and wordy'. The designer, trained to coordinate words, hands and visual imagery, is still very much the exception.

The skill of visualization is most frequently used by the industrial designer to synthesize other people's ideas, and in particular to provide concreteness to marketing and engineering concepts. In 1955 Henry Dreyfuss wrote that the designer

can sit at a table and listen to executives, engineers, production and advertising men throw off suggestions and [can] quickly incorporate them into a sketch that crystallizes their ideas – or shows their impracticability.[25]

But the skill can also be used to pull together the designer's own ideas – if only they are not suppressed in early life by peer group pressure from other functions, and by the corporate hierarchy.

In some of the companies where industrial designers have been allowed to play a full part, they are not only supplying their own

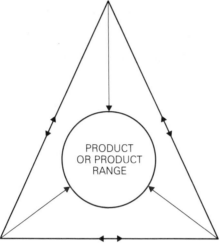

INDUSTRIAL DESIGN VISION

● aesthetic knowledge (form, colours)
● social backgrounds, cultural backgrounds
● environmental relationships
● ergonomic requirements
● visual trends
● insight into aspects of marketing & engineering vision

PRODUCT
OR PRODUCT
RANGE

MARKETING VISION

● market research
● market analysis
● economics
● distribution systems

ENGINEERING VISION
(DEVELOPMENT &
PRODUCTION)

● technical research
● technical analysis
● economic targets
● production methods
● ergonomic research

Figure 1 *The ideal pooling of cross-functional 'vision'; but the industrial designer can also supply marketing and engineering vision (adapted from Concern Industrial Design Center, Philips, Eindhoven).*

brand of vision, but are effectively substituting for the vision which ought to be part of the armoury of marketing and engineering, but which is often so signally lacking. At the very least, they are acting as catalysts in the development of a common product imagination for the management team.

This concept is given graphic form in figure 1, which is based on a diagram in widespread use within Philips. As the case studies in part II make clear, the well-rounded industrial designer has much to contribute in respect of almost every aspect of the vision which engineering and marketing rightfully ought to exercise: from ergonomics and the design of new production methods (on the engineering 'side of the house'), to new ways of analysing markets and conducting (or interpreting) market research. It is not the mundane skills of sketching, shaping and colouring which make the industrial designer such a valuable resource, but the multifaceted ability to contribute to the work of other disciplines, and to stimulate, interpret and synthesize it. A company does itself a disservice if it sees product design, and with it the industrial designer's contribution, as merely 'shape and appearance'. Yet that is the rationale which is still often put forward for the renewed interest in design that is developing among businesses across the world.

In late 1984 *Business Week* reported that

in a world where many new products are similar in function, components and even performance, a product's design – its shape, its look, and above all its image – can make all the difference.[26]

In so far as the shape and look of a product affect the more obviously 'functional' aspects of its design, this doctrine may very well constitute an argument for starting to involve the industrial designers very early in the product development process – as they have been in a handful of European companies since the 1920s. But it is still perilously close to Harley Earl's strategy of skin deep styling, and is inadequate in the modern world. Except in products where function is of little significance, and emotional make-believe almost all, this approach is unlikely to pass Levitt's test of meaningful distinction.

As the new corporate design converts have learned in recent years, it is in helping to achieve *real* differentiation that industrial design can play such a valuable part. To do that, the industrial designer needs to be able to work from the outside-in, in contrast with the engineering designer's tendency to work, through layer upon layer of technical detail, from the inside-out. Rather than just tinkering around with the product's wrapping, the industrial designer must start with the complete product as it would be used by someone, and then go back into the details required to make the concept work.[27] In order to do this, the industrial designer must be a fully-fledged member of the management team, with all the power and 'muscle' which that implies.

3 The Marketing Conundrum:
A Curious Case of Persistent Myopia

Recognizing a good idea is easy. Implementing it is quite another matter, especially if it seems futuristic, or if a full-blooded revolution is required. Unless change is clearly a matter of life or death, it will tend to be suppressed.

Marketing is no exception. The notion that corporate survival depends on doing everything necessary to satisfy the needs and wants of the customer was hailed as a brilliant breakthrough in 1960 when it was popularized by Theodore Levitt's manifesto-like article 'Marketing Myopia'[28] after the pioneering efforts of Procter & Gamble some years earlier.

For a time it looked as if the dogma of 'know thy customer' would sweep the world. Companies rushed to install vice-presidents and directors of marketing, and to beef up their market research. The question 'what business are we in?', which Levitt promulgated as the necessary foundation of a marketing-led strategy, quickly sprang to everyone's lips, and encouraged a veritable passion for corporate analysis.

The 'marketing concept', as it has come to be known, holds that the objective of every company in an age of abundance should be to focus, in a myriad of ways, on the customer's present and future needs and wants, thereby winning his or her satisfaction and loyalty. Among the techniques the company should employ are market segmentation, product differentiation, and a careful blending of the various elements of the 'marketing mix', including product features (both function and appearance), price, promotion, distribution and after-sales service.

Marketing gurus with manifestos at the ready: Theodore Levitt (left) and Philip Kotler.

Expressed in Levitt's forthright words,

selling concerns itself with the tricks and techniques of getting people to exchange their cash for your product. It is not concerned with the values that the exchange is all about. And it does not, as marketing invariably does, view the entire business process as consisting of a tightly intregrated effort to discover, arouse and satisfy consumer needs.[29]

Yet for most of the past quarter-century it has only been in 'packaged goods' like food, drink, cosmetics and soap powder that marketing has been practised by an entire industry in anything near the breadth and depth that Levitt advocated. Even there, it has degenerated in some leading companies into a narrow, functional activity. In other industries, the Japanese quickly learned to apply a marketing approach to everything from cameras to cars, from consumer electronics to copying machines.[30] But most Western makers of such consumer durables still suffer, in common with the suppliers of capital goods and services, from what Levitt branded as a traditional short-term production/sales mentality. They may *think* they are practising marketing, but they have confused the tools with the concept. They have merely been doing market research, advertising, promotion and sales – and sometimes only one or two of them.

Exceptions have abounded for years: Black & Decker in consumer durables, IBM in capital goods (and, via its personal computer, in durables too), McDonald's in services. But it was only after the start of the 1980s that Ford, Philips, Apple, Hewlett-Packard and a host of companies in all sorts of industries – even chemicals, insurance and banking – sensibly rushed to join them in creating what has been called a 'new era of marketing'.[31]

In general, as Tom Peters and Bob Waterman wrote in *In Search of Excellence*, it is still true that 'despite all the lip service given to market orientation these days . . . the customer is either ignored or considered a bloody nuisance'.[32] Most managers continue to see their company's essential purpose as selling whatever production happens to make, rather than designing new products and services to suit the changing preferences of the customer.

Peters and Waterman had plenty of evidence for their assertion. In Europe, some companies have not even defined what markets they are in, or who their competitors are. Not only do they have precious little understanding of marketing as an overall concept, but even some of its basic techniques – such as segmentation and differentiation – are still foreign to them. Where companies do make conscious use of the 'marketing mix', they tend to emphasize only one or two of its elements: price and promotion, say, or distribution and after-sales service. They seldom exercise Japanese-style control of the entire mix. And more often than not the first of marketing's 'Ps' – the product – gets short shrift.[33, 34] Hence, in part, the neglect and misuse of industrial design.

Surprisingly, given the usual assertion of a yawning management gap between Europe and the US, marketing seems almost as under-developed in many American companies. In Philip Kotler's words,

we've been talking about marketing for 25 years, but very few companies really do it. A lot of chief executives are confused about the difference between marketing and sales. They don't seem to realize that most of the impact of marketing is felt *before* the product is produced, not after.[35]

Which is not, of course, to say that marketing consists essentially of market research – a source of dangerous confusion about which industrial designers have particular cause to complain. Without proper marketing thinking it is impossible to get market researchers to ask the right questions. And it is just as hard to make effective use of its results. Market research in isolation imposes a stranglehold on corporate imagination and creativity.

The Concept (and how the Japanese Apply it)

There was certainly no confusion about the meaning of marketing in the minds of the various management thinkers who, in the 1950s, laid the foundation for Levitt's popularizing manifesto. Peter Drucker, Wroe Alderson and the other formulators of the marketing concept avoided a slavish attitude to market research, as did those who were credited with the ground-breaking thought on segmentation and differentiation.[36] So did Procter & Gamble, the company credited with pioneering the concept in practice as early as the 1940s.

'Marketing Myopia' itself was quite clear on the subject. Though the American automobile industry had for a long time done lots of consumer research, Levitt wrote in 1960, the success of the Volkswagen and other imported 'compact' (small) cars showed that it was failing to reveal what the customer really wanted. 'Detroit never really researched the customer's wants. It only researched his preferences between the kinds of things which it had already decided to offer him. For Detroit is mainly product-oriented, not customer-oriented.'[37]

This attitude of 'put the company before the customer' persisted for another 20 years, until Detroit (or, at least, Ford) decided that the time had at last come to give priority to the consumer.

In contrast with Detroit, Levitt wrote,

a truly marketing-minded firm tries to create value-satisfying goods and services that consumers will want to buy. What it offers for sale includes not only the generic product or service, but also how it is made available to the customer, in what form, when, under what conditions, and at what terms of trade. Most important, what it offers for sale is determined not by the seller but by the buyer.[38]

As Philip Kotler pointed out in his definitive book on Japanese marketing strategies, *The New Competition*, this was a lesson the Japanese had already begun to learn by the early 1960s.[39] After the dismal failure in 1958 of the seriously flawed Toyopet, Toyota's first attempt to export cars to the US, the company quickly adopted a fully-fledged marketing approach.

Through assiduous market research, it discovered that the market segments already served by imports of Volkswagen 'Beetle' were growing, and also that the way to beat VW was to improve on the German company's already-high standards of product quality and performance, driver and passenger comfort, and after-sales service.

In addition to this meticulously precise approach to market definition, differentiation and product development, Toyota applied carefully pitched strategies on pricing, distribution and promotion. The Toyota Corona was a resounding success, as was the next major car the company introduced, the Corolla.

But Toyota did not stop there. Not only did it pursue a policy of continued product improvement, but it steadily refined its marketing strategy still further. Like most of the Japanese companies which have been successful in world markets since the 1960s, Toyota did not stick blindly with one approach.

Emulating the Samurai warriors of old Japan, it learned not one martial art, but several, so that it could choose the best means of attack or defence in any particular situation.

As *The New Competition* makes abundantly clear, Japanese companies sometimes attack with a karate blow aimed at a competitor's single weak spot, sometimes with an aikido side-step and sometimes with a full-frontal judo throw. As they enter, penetrate and finally build leadership in their carefully chosen markets, they are adept at wielding the different weapons of the marketing mix with carefully varied degrees of emphasis. In any one of several combinations, they use them to flank, encircle or bypass the enemy, or to mount a frontal attack.[40] Above all, they are remarkably persistent.

The Japanese are particularly adept at gradually 'cascading' from highly focused, small market segments into related, larger ones. This approach has been followed with precision in consumer electronics, starting with transistor radios in the 1950s and gradually broadening through tape recorders, monochrome TV, colour TV and video cassette recorders (VCRs) (see appendix).

The focus adopted by Japanese companies in order to obtain market entry is extremely tight. It concentrates not just on carefully selected customer types in particular regions, using hand-picked dealers and distributors, but often on specific towns or parts of cities.[41]

Once the market has been penetrated, a range of product development strategies is applied. One is 'product line stretching' – broadening the line in order to reach a wider segment of the total market. Another is 'product proliferation' – the introduction of a multiplicity of product types or models at each point in the product line. This not only allows the Japanese company to appeal to a large number of market niches, but also ties up distribution channels and retail stores, depriving competitors of space. There are countless examples of this method of attack: Sharp and Casio in calculators, Seiko in watches, Canon in cameras, Honda in motorcycles, Toyota and Nissan in cars.[42]

The Japanese are also masters of pricing. In every target market they have entered, they have applied so-called 'market-share pricing', deliberately using a low entry price to build up market share and establish a dominant position. Lower prices not only allow them to promote their products as offering greater value for money, they also help them move down the 'experience curve' of high factory output, thereby lowering production costs still further. Pricing is a strategic weapon, in other words – just like everything else in Japanese marketing.

The Pitfalls (and how the West is Caught)

So why, despite the obvious good sense of Levitt's manifesto, and the mind-concentrating example of marketing-led Japanese competition, did so many Western manufacturers fail to convert from selling to marketing?

In itself, there is nothing remarkable about some industries taking longer than others to become marketing-minded. If a maker of machine tools could sustain a sufficient competitive advantage through high technology, low production cost and an efficient sales and service network, there was little point trying to add a fully-fledged marketing strategy to his armoury. Hence, for example, the fact that Apple and Hewlett-Packard waited until the mid-1980s before embracing marketing.

Similarly, when a service industry such as banking or telecommunications was so regulated as to suffer only minimal competition, it was understandable that marketing should be given low priority, however annoying this might be to the customer.

Yet many companies should have moved much more quickly. Especially in cars, consumer electronics and other consumer durables, where Japanese marketing strategy has been so obviously lethal for so long, they have taken a remarkably long time to respond to competitive example.

A fundamental part of the explanation for their dilatory behaviour was foreshadowed by Levitt himself in 'Marketing Myopia'. Building an effective customer-oriented company, he wrote, 'involves far more than good intentions or promotional tricks; it involves profound matters of human organization and leadership'.

Not only must top management show leadership, vision, the will to succeed and the ability to motivate, but

the entire corporation must be viewed as a customer-creating and customer-satisfying organism. Management must think of itself not as producing products but as providing customer-creating satisfactions. It must push this idea (and everything it means and requires) into every nook and cranny of the organization. It has to do this continuously and with the kind of flair that excites and stimulates the people in it.[43]

In other words, the corporation must not only undergo a radical shake-up in organization, but also a complete cultural revolution. In particular, it must embrace the reality of institutionalized subversion. For if a company is to respond to changing consumer preferences, and to competitive actions, it must constantly question its existing strategies and tactics. Its keynote must be flexibility. Yet constant change is anathema to most corporations. Faced with a challenge of such magnitude, it is not altogether surprising that so many managements have either funked it completely, or have failed to appreciate all the pitfalls.

The innumerable reasons for failure can be summarized under a dozen headings. They are complex and interactive, and exacerbate each other. Any order of ranking is therefore arbitrary; the opening duo, accorded that status because of their particularly pernicious impact on the use – or neglect – of industrial design, result from several of the factors lower down the list.

(1) Confusion with sales

This is one of the most common errors of all. It occurs for several reasons. One is that the concept of marketing is harder to understand than are its various tools. Another is the way the word 'marketing' was quickly adopted by sales and promotion as a grander term for their own activities. A third results from the tendency to promote sales managers to marketing positions. As Kotler comments

> the new marketing executive continues to think like a sales executive. Instead of taking time to analyse environmental changes, new consumer needs, competitive challenges, and new strategies for corporate growth, he spends his time worrying about the disappointing sales in Kansas City last week, or the price cut initiated by a rival corporation yesterday. He is probably involved almost as much in putting out new fires as he was when he was a sales executive.[44]

Whereas the sales executive's orientation is towards individual customers, and is essentially short-term (see point 9), the attention of the marketing executive should be directed at long-term trends, threats and opportunities, and at market segments.

(2) **Confusion with market research, and domination by it**

Many managers are heavily oversold on what market research can do. To repeat what has become a cliché in the marketing world, they use it as a decision substitute, instead of for decision support.

This is highly dangerous on several grounds. The predictive abilities of market research are innately limited, because it can only probe attitudes, which are often a surprisingly poor guide to actual behaviour. This applies particularly to the sort of quantitative research carried out through mass questionnaires and consumer testing (the users of such methods are known by market researchers as 'nosecounters'). But it is also true of much of the creative (or 'qualitative') research conducted with the help of psychological techniques by experts familiarly known as 'headshrinkers'. Even if the research is well and objectively constructed – and there are plenty of horror stories of loaded questions, poor sampling techniques and other methodological errors – market research can probe the human psyche to only a limited degree of reliability.

One of the most devastating critics of the misuse of market research is Stephen King, a long-time director of J Walter Thompson, the big international advertising agency, and one of the profession's leading intellectual 'consciences'. In an analysis which would be hilarious if it were not so depressing, he attacks the vast entrail-gazing industry of market research for being 'dedicated to taking informed guesses to four places of decimals'. Worse, he alleges that

marketing men are looking for simple answers about the motivations for buying their brands, because they want market research to tell them what to do. They are hoping for a neat, rank-ordered list of motivations so that the top half-dozen can be stuffed directly into products and advertising.[45]

The failure of many companies to innovate, maintains King, has resulted from an all-too-common use of market research as a system that ideally reduces all personal judgement to a decision as to which of two numbers is the larger. In such a climate, it is well-nigh impossible for marketing to apply whatever imagination it may possess. Obsessive reliance on market research, and a mistaken belief in it as the provider of direct answers, completely stifles any attempt to create that 'meaningful distinction' which is at the heart of competitiveness in today's crowded marketplace.

The fundamental point is that most market research techniques merely measure what the consumers themselves know they want or prefer. This may be fine when a company is deciding whether to change the colour of an existing product, to improve its reliability, or to add a new feature. To a considerable extent this is what Toyota did in the US in the early 1960s, when it carefully analysed how it could improve on the way Volkswagen was penetrating the market. But market research was only one of the inputs to its decision-making process; judgement was the other.

The Western motor industry's traditionally slavish reliance on consumer 'clinics', and other forms of testing, is part of the explanation for its failure to innovate in any real sense between the 1930s and the 1970s; the story of how Ford broke with this blinkered tradition in the late 1970s, with dramatic results, forms part of chapter 8. It was General Electric's equally narrow dependence on consumer testing which convinced it that there was no market for small-screen monochrome televisions.

In June 1960 GE had commissioned what by any standards was a major piece of market research. Using mock-ups of black and white television sets with eight different screen sizes, weights and prices, it set out to estimate the potential demand for a portable set with a small screen. After hours of interviews, the study came up with several negative conclusions, including that 'people do not place a high value on portability of the television set'.

Just a few weeks later Sony unveiled a tiny 8-inch receiver at the Chicago Music Show. It went on mass sale a year later as a luxury item, priced at $250 when discount stores were selling middle-of-the-line 21-inch sets for under $150. Promoted by a highly imaginative advertising campaign and sold by Sony's US sales subsidiary direct to department stores and other large retailers – by-passing much of the sceptical trade – the product was so successful that other Japanese companies rushed in with their own versions.

From a negligible level in 1960, Japanese TV imports to the US soared to 120,000 in 1962 and a million in 1965. By almost completely cornering the market for screen sizes of under 12 inches, the Japanese had established a base from which they could expand to larger monochrome units, then into colour and eventually into video recorders.[46]

Rather than asking consumers to predict their feelings towards a type of product which was unfamiliar, what Sony had done was to observe the two factors which together would soon change the pattern of TV viewing, and create a demand for a second set in many homes: the rapidly increasing penetration of large TV sets into the American lounge and the growing proliferation of TV channels. In effect, the company had looked beyond consumers' expressed needs to their underlying behaviour patterns, and had led the market by stimulating a new want. It has since done precisely the same with the video cassette recorder, the Walkman personal stereo and the Watchman flat-tube TV.

In other words, though senior Sony executives have been heard to declare 'I don't believe in market research – it doesn't help us develop new products', the company actually does make heavy use of market research – not the conventional kind, but an amalgam of fact-finding and social forecasting. And it gives an equally influential role to the sort of informal observation and intuition exercised so successfully by its founder Akio Morita, and practised today by its industrial designers. Other companies have much to learn from its example (see chapter 7).

(3) **Confusion with customer service**

This is otherwise known as the 'Have a nice day' syndrome. Many companies, especially airlines, banks and other service organizations, seem to think that marketing essentially consists of smiling at people and uttering meaningless catch phrases. Like those who confuse marketing with sales, market research, pricing, promotion or any other single element of the marketing mix, they are treating marketing merely as a set of tools.

(4) **Organizational barriers**

The creation and implementation of marketing strategies runs foul of barriers between the corporate centre and line management, and between different functional specialists. Much of the success of Japanese management lies in its ability to cross disciplinary lines. But this 'group process' is totally alien to many companies in the West. Managers see themselves as individuals and specialists, rather than as members of a team. This has encouraged marketing itself to be seen as a specialist function, rather than as a diagnostic way of thinking which needs to permeate the entire company. This tendency is reinforced when specialists from sales, production and finance are moved into marketing jobs.

(5) **Structural failings**

In many Western companies the problem of poor communications and inadequate cohesion is compounded by a failure to break up the organization into smaller business units, with 'product managers' responsible for a handful of products and/or markets. Without such a structure it is difficult to get sufficiently 'close to the customer', to permeate each part of the organization with a driving commitment to the performance of each product line, and to create that elusive collaboration between different functional disciplines.

(6) Hostile corporate cultures

Outside packaged goods, most companies are dominated by a technical, sales or financial culture. In 'Marketing Myopia' Levitt warned that executives with such backgrounds have an almost trained incapacity to see that achieving volume may require them to understand and serve many discrete market segments, which may each be quite small. Instead, they insist on trying to serve an elusive, often mythical, mass market. Such executives also have an inbred tendency to think short-term, rather than with the long-term horizon which much marketing strategy requires.[47] In such a hostile environment, attempts to introduce marketing tend to fall on stony ground, and marketers are thought of as fast-talking upstarts.

(7) Myopic marketers

Marketing people need to be strategic thinkers and exceptionally good communicators, especially if they are brighter and brasher than their peers in other functions. But their training and work experience tends to turn them into over-numerate technocrats. By no means all of them receive a formal training; for those who do, it is techniques which are stressed, not human behaviour, motivation and communication.

(8) The strategic planning barrier

The rapid development of strategic planning during the 1970s eroded the status and role of marketing strategy, encouraging still further the view that marketing is merely a set of operational tools for use only by line management.[48] Strategic planning concentrated on financial and economic analysis (especially of costs and cash flow), and paid little attention to marketplace factors, other than stereotyped descriptions of market growth and market share.

(9) **Short-term horizons and lack of persistence**

By bolstering the influence of the finance function, strategic planning reinforced the innate short-term bias of many Western companies, especially in the US and Britain. Strategy became dominated by financial rules of thumb, such as Return on Investment (RoI), which operate with a very short horizon. Such measurement systems can clash head-on with strategic marketing.[49]

This point cannot be made forcibly enough. In *The New Competition* Kotler and his co-authors rightly warned that 'if you seek a two-year payback period, it is difficult to go to battle with competitors who disregard short-term returns' – as the Japanese so manifestly do in many product markets.

The most dramatic illustration of how persistence pays is the video recorder. All the companies which developed VCR technology in the 1960s and 1970s tried prematurely to commercialize it in the form of a consumer product, and failed at least once. In 1973 Matsushita geared up an entire department with 1200 employees to produce a VCR that totally failed in the market. And the Betamax, which Sony launched in 1975 after a near-20-year development programme, was the fourth generation of VCR it had demonstrated as a consumer product.[50]

While Sony and Matsushita remained committed in the face of disappointment and failure, along with Philips, the several American companies which tried to break into the technology quickly withdrew after the failure of their early efforts.[51] In other sectors, too, this was the essence of America's failure to respond to the Japanese import invasion of the 1960s and 1970s. The vast majority of companies continued to take the easy option of merely exploiting today's opportunities, rather than also preparing for tomorrow's. Only in the 1980s is this lesson gradually being learned, as more US and European companies try to emulate the Japanese strategy of long-term product/market development.

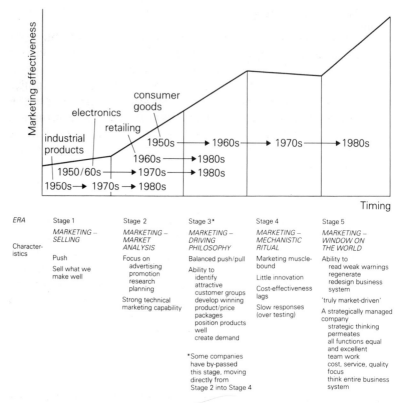

Figure 2 *The evolution of marketing effectiveness.*
(Source: McKinsey & Co).

(10) **The 'theory gap'**

In parallel with recent self-criticism by corporate strategy theorists, marketing academics now admit that marketing theory has not helped its own case by paying inadequate attention to the organizational and human aspects of how strategies can be implemented.[52] When not dealing with grand strategy, marketing teaching has tended to be obsessed with the detailed techniques of pricing and market research; these may be the stuff of which academic careers are made, but on their own they are not sufficient to tell a company how to implement a strategy of, say, penetrating a premium market segment.

(11) **The 'competition gap'**

Though the marketing concept itself was born out of the need for companies to compete more effectively, it was taught (and practised) as if the satisfaction of consumer needs was, on its own, a sufficient criterion for achieving profitability. Its clear message was 'know thy customer', not 'know thy competitor'. Business seemed to be a race which all could win.[53] Now that competitive strategy has become such a crucial element in business survival, marketing academics are rushing to fill this gap.

(12) **Degeneration**

Successful past practitioners of marketing have become inflexible in their approach, and therefore slow to respond to changes in competitive forces, market structures and distribution channels. A prime example is provided by the problems suffered by packaged goods manufacturers in Europe since the late 1970s, as a result of the strong rise in retailers' own brands. This sort of inflexibility has been christened 'muscle-bound marketing'.

Conclusion

Ideally, marketing should be one of the main driving forces of a company – any company. It should pervade the corporate culture and transcend any functional department or title. In practice, it tends to take root gradually (see figure 2). It may become increasingly deep-rooted and effective, only to degenerate into a mechanistic ritual (this may occur immediately after stage 2 in the diagram, without it ever having become a driving philosophy). Yet only if marketing reaches stage 5, and really becomes the company's 'window on the world', can it be fully effective. This is the route the Japanese have pioneered, and that the West is belatedly beginning to follow.

4 Towards a New Era:
Design, Marketing and Technology as Equal Partners

Hewlett-Packard has always been universally acclaimed as one of America's best-managed companies. But in the summer of 1984 it sent shock waves through its engineer-dominated ranks by establishing a centralized marketing division and making other controversial changes to its structure – all of them aimed at a radical improvement to its marketing.[54]

Like Apple the previous year, HP had come to the uncomfortable realization that it could no longer compete just with technology: it had to become more responsive to the preferences of the customer, more choosy about which market segments it should serve and more effective in its selling and distribution.

Since the early 1980s countless other Western makers of consumer durables and capital goods have started to undergo the same long conversion. And with it has come – or is coming – a dramatic increase in the influence of industrial design. For as Apple discovered quite early in the process[55] (along with European companies such as Philips – chapter 9), many of the factors which demand new priority for marketing also require an upgrading of design. Given the dedication to design of arch-competitors such as IBM, Olivetti and Apple, it can only be a matter of time before Hewlett-Packard and co follow suit.

Behind the elevation of marketing lie several categories of change: in management resources; in international economics; in product and process technologies; in communications technology; in society at large; in distribution and retailing; in government regulation of industries; and, above all, in patterns of competition.

The main change in management resources has been a growth in sophistication, particularly towards corporate strategy and marketing. This has resulted in part from influences internal to the business world. Executives who trained at business schools or similar institutions have gradually come to occupy most of the positions of power – almost universally in the US, and to a growing extent in Europe. The widespread introduction of decentralized 'strategic business units' during the 1970s has pushed strategic responsibility well down the organization. As a result of both these trends, what might be called a 'strategic mentality' has begun to permeate many organizations, even at low levels. It can handle complex analysis and planning, and despite the problems discussed in chapter 3 is beginning to produce results, in the form of a more expert development and execution of complex marketing strategies.

The other key influences behind the rise of marketing are external to the company itself. (For simplicity's sake, I have only selected factors which affect markets in the industrialized world.)

For some industries, such as banking, financial services and telecommunications, it is changes in communications technology, together with deregulation – and therefore extra competition – which have had the most impact. For packaged goods makers, it is a combination of all the factors (in Europe, even deregulation has been significant: the abolition of retail price maintenance has shifted the balance of power from manufacturer to retailer, forcing the former to develop more imaginative marketing strategies).

For the manufacturers of consumer durables and industrial products, who constitute the main focus of this book, the main influences are economic, technological (all types), social and competitive. As with the catalogue of marketing pitfalls in the last chapter, they are complex and interactive. But in this case they constitute a list of only ten.

(1) The slowdown in international economic growth

This began in the early 1970s, but was a particularly significant factor after the second OPEC 'oil shock' of 1979; it has been especially persistent in Europe. This factor has deeply affected most of the other items in this list.

(2) **Product/market maturity**

Certain markets have matured as penetration reached ever-higher levels; obvious examples include cars, hi-fi, television, typewriters, washing machines. To design products which will sell profitably in such a marketplace requires marketing and design imagination of the highest order.

(3) **The splintering of markets**

This partly results from social trends (more single-parent families, higher unemployment, smaller dwellings, etc.). But it is also the result of conscious competitive strategies, especially on the part of the Japanese. It creates the need for meticulous target-marketing and design.

(4) **Competition**

There has been an intensification of competition from every quarter: from domestic and foreign suppliers, from established industry competitors and new entrants.

(5) **Developing-country exports**

There has been a particularly sharp growth in exports from the 'newly industrializing countries'. Since the early 1970s this has made it almost impossible for a Western manufacturer to compete on price alone; now Hong Kong, Singapore, South Korea, Taiwan and other countries have started to compete on quality as well. Again, this demands extremely imaginative marketing and design.

(6) **Technological maturity and the pervasiveness of new technology**

Technology no longer remains proprietary to one country or company for long, whether it be the latest silicon chip, a new type of plastic or the latest type of robot for the production line. This has two main effects:

(7) **It is becoming increasingly difficult** to sustain a competitive advantage on the basis of technology alone;

(8) **Product life cycles are shortening.** This trend is being accentuated still further by the competitive strategies of many companies, especially the Japanese.

(9) **Shorter development cycles**

Partly because of shorter product life cycles, but also as a response to the spiralling costs of research and development, companies are striving to shorten the cycle between the start of a product's development and its launch into the market.[56] Among many other changes, they are having to streamline their traditionally protracted procedures for technical and market testing. Procter & Gamble, the giant American packaged goods multinational, used to take up to two years to market test each of its new products, usually on two million consumers (1 per cent of the US population), and in several major regions. But in many cases it now uses improved methods to simulate such tests in small samples, sometimes of only 2000 people, so that it can confine marketplace testing to a couple of cities. Several products launched since the early 1980s were tested for barely a year, including 'Ivory' shampoo and 'Citrus Hill' orange juice.

On the technical side P & G's tactics included starting work on new production plant at the same time as market testing of the product, rather than waiting until it was virtually completed and then taking another four or five years. In the case of Citrus Hill, launched in 1983, this more than halved the development-to-production cycle to less than four years.

On a wide range of products, P & G also introduced more rapid environmental tests. A few miles from its headquarters in Cincinnati, on the banks of the Ohio River, its Ivorydale detergent factory housed by the mid-1980s a small-scale river, complete with an entire ecological chain of plants and bugs, plus a water treatment installation. Bringing the real world into the laboratory in this way cut the time needed for some tests from 18 months to just six.

(10) **Globalization**

With the exception of items (1) and (2), all the above have been intensified by the emergence of 'globalization', whether

The ideally balanced partnership between design, marketing and technology.

in the form of global competitors who vary their product offerings from one market to another, or in the more concrete shape of global products and even (in some cases) global brands. This trend is examined in part III.

Not only is the formulation of global strategy a major challenge in itself – for marketing, design and every other managerial

function – but its execution requires far-reaching changes to be made in corporate structures and procedures. This is particularly true of the product development process. It can no longer be a leisurely affair in which the project is cautiously passed from one department to another in a series of sequential phases. Instead, the phases have to be made to overlap, and marketers, product planners, industrial designers, development engineers, production engineers and all the other functional specialists have to work in parallel, in a tightly integrated team, and under intense time pressure.

This approach was pioneered by the Japanese[57] and has been ardently copied by a growing number of Western companies, in all sorts of industries, from cars to computers, and food to consumer electronics. Apart from P & G, an already legendary example is provided by IBM, which with its first foray into personal computers beat the 'PC' industry's 24-month development norm by almost 10 months, and its own previous time-scale on bigger machines by almost three years. Apple and IBM's other competitors were forced to try to match this new 14-month norm. The consumer electronics industry is peppered with similar examples, even in such traditionally slow-moving companies as Philips (see chapter 9).

In particular, this method of development involves better design-for-manufacture, as well as the tactic adopted by P & G of starting much earlier with the design of production plant and the build-up of manufacture.

One of the most illuminating studies of accelerated product development looked at comparable Japanese and American high-tech companies. It found that the Japanese used a longer initial design phase, the second half of which was run in parallel with much of the advanced development phase. As a result, few design decisions had to be reworked, and the manufacturing build-up phase could be drastically shortened. The net effect was that the entire Japanese design-to-production cycle lasted only 20 months against the American 30.[58]

In these circumstances, where interdepartmental friction, time, cost and risk must all be minimized, the synthesizing, integrating and communicating skills of the industrial designer are becoming especially valuable. But the designer can also play a central part in helping the company take the sort of risks which are increasingly imperative in today's ultracompetitive business environment. Here it is the industrial designer's imagination, as well as the skill to make connections, which is of paramount importance. The connections may be between new technology and an existing consumer need, as in the design of a small, lightweight video camera recorder. Or they may be between well-established technology and unlikely new patterns of consumer behaviour, as in several of Sony's money-spinning products. Whether as a substitute for marketing vision, or as a key part of a properly managed marketing strategy, the design imagination has become a vital competitive weapon in the struggle to satisfy the customer and beat the competition.

Part II
Design Unchained

Introduction

Industrial design has developed commercial 'clout' in many different ways. In some companies it has struggled for years to make its presence felt, only gradually winning a place in executives' hearts and minds. In others it has suddenly been upgraded in response to a dramatic change in the company's circumstances, such as the failure of its established marketing strategy, the urgent need to stimulate new product development or the arrival of a new chief executive; often the three go hand-in-hand.

By the same token, corporate 'conversions' to design take various forms. Some companies give priority to product design, others go from the start for the full design spectrum, also developing better environments and visual communications. Many use design consultants, others rapidly construct a strong in-house department; some departments are centralized, others are diffused across the organization. Some companies see industrial design as a partner of marketing, others as a partial substitute. A few openly recognize its 'bridging' role between marketing and engineering specialists, many play this by stealth.

The choice of route depends not only on the company's situation, but also on its culture. Some companies are dominated by marketing experts, others by development engineers, still others by production, sales or finance. Many readily recognize their shortcomings, others are reluctant to do so. The only common denominators between the new design converts are the need to strengthen design and give it an independent voice; the importance of top management commitment; and the pivotal role played by a few key personalities in creating the necessary change of attitudes and priorities.

These contrasts and commonalities are illustrated in depth in the case studies of 'design unchained' which form the next seven chapters. Each chapter describes the changes in strategy which have prompted the various companies to upgrade industrial design and to reinforce their new-found commitment to it. The changes are described not only in general terms, but also through the story of specific product development projects.

The first two companies, Olivetti and Deere, are long-standing pioneers of design-minded management, but they came to design in very different ways: Olivetti almost from birth, Deere only gradually. Though they are both committed to excellence right across the design spectrum, in other ways they have adopted dissimilar approaches to design management.

Four more recent converts are then examined. Sony's conversion is particularly rich in lessons for other companies, including: the need for a strong and independent design function even if the company's marketing is unusually skilful and imaginative; the way designers have to fight for their corner, even when championed by top management; and the pros and cons of centralized design departments (Sony changed the structure of its design unit three times between 1978 and 1985).

Ford and Philips share many similarities. Both were dominated by technical cultures – Ford by production, Philips by science and engineering. So in order to execute their new corporate strategies, both companies have had to mount a marketing revolution as well as a self-induced design conversion. With marketing imagination thin on the ground, designers in both companies have had plenty of scope for seizing the initiative – and they have taken it. The Ford chapter focuses especially on the gradual unfolding of the company's design-led marketing strategy, first in Europe, then in the US. It gives special emphasis to the changes that have had to be introduced in corporate structure and procedures, including in market research, to make the twin revolution a success. The chapter on Philips deals more briefly with the commercial background, concentrating on the very radical changes which have been made in the way the company manages design and development. The chapter on Baker Perkins, the unlikeliest convert of all, describes the company's gradual discovery of the value of industrial design skills, and recounts in some detail the interplay between design, marketing and engineering in the management of a major development project.

Finally, the chapter on the work of Kenneth Grange offers a wide range of insights into how different companies handle the complicated business of using a top international design consultant to create new products – and sometimes new strategies. In his everchanging role as initiator, synthesizer, coordinator and catalyst, Grange epitomizes the new-found strengths of industrial design which permeate the design converts.

5 Confronting a New Challenge: Olivetti

Olivetti has always prided itself on setting trends which much larger companies are then forced to follow. Having burst into IBM's supposedly impregnable territory in the late 1970s with the world's first fully electronic typewriter, it immediately took the lead with the slimmest and most 'user-friendly' computer keyboard.

The success of these products, plus its position of international leadership in computer ergonomics, played a key part in the Italian company's transformation in the early 1980s from elegant but unprofitable duckling into attractive, healthy swan. By the time American Telephone & Telegraph, the giant telecommunications company which has become engaged in an international duel with IBM, came hunting for a European partner in office electronics, Olivetti had become the apparently ideal candidate. Just before Christmas 1983, AT & T bought a 25 per cent shareholding – and Olivetti became part of a global enterprise. The fact that the alliance later languished had nothing to do with any American disappointment at Olivetti's innovative culture. Rather, it was caused by a clash of cultures, both corporate and national: Italian entrepreneurs and American machine bureaucracies do not work comfortably together.

Without Olivetti's world-renowned industrial designers, neither the typewriter nor the computer keyboard would have made much of a mark. It was Mario Bellini and his staff, working closely with the company's engineers, who played the leading role in applying some highly advanced electronic circuitry and 'creating' the

International pace-setters. Top: Olivetti's 'ET' electronic typewriter range, 1978. Designer: M. Bellini. Bottom: Olivetti's 'L1' keyboard, 1980. Designers: Sotsass, Sowden.

easy-to-use, imposingly wedge-shaped black 'ET' range of type-writers – an image which the Japanese and other competitors tried to copy. And had it not been for Ettore Sotsass' team, the 'L1' keyboard might never have been born – certainly not as the first of its kind, giving the company an invaluable competitive lead, and acting as a lever for the sale of millions of dollars worth of computer systems.

The two machines epitomize Olivetti's long-standing ability to enhance the value of its products in the marketplace through the use of industrial design. Its entire range of computers, terminals, typewriters, screens, printers and so on are not only sleekly styled, but also possess all sorts of features designed to make them easy to use and convenient to operate. From the slim, adjustable keyboard, to antiglare screens on retractable arms and all sorts of other aspects of operator comfort, Olivetti has become a world leader in 'user-friendly' hardware, repeatedly giving itself a handsome competitive edge.

Its commitment to excellent and striking design extends throughout the workplace, to furniture and architecture, as well as to products and systems. This dedication also permeates all the conventional techniques of corporate self-promotion: graphics in every piece of documentation, in exhibitions, in advertising, and the frequent sponsorship of art exhibitions, for which the company is renowned throughout the world. In all aspects of its activities, not just for the occasional flagship product or for the furthering of its corporate image, Olivetti uses design to the full. Not for nothing has it been one of the very best models in the world of a thoroughly 'designerly' firm.

Behind this far-reaching commitment lies more than half a century of top management belief that design should be one of the key factors in corporate strategy and marketing, and that designers should therefore be given a pervasive influence at all levels of management, and from start to finish in the product development process.

The company was founded in 1908 by Camillo Olivetti, an engineer and measuring instruments maker who had just spent a seminal period working in the US. As the location for his new headquarters, and for his first typewriter factory, Camillo chose his native Ivrea, a small town which nestles in the foothills of the Alps north of Turin.

Though strongly influenced by his experience in the US, Camillo had typically European ideas about the appearance of products. His first typewriter, introduced in 1911, looked more functional than its American competitors. Camillo wrote at the time that

a typewriter should not be a gewgaw [gaudy trifle] for the drawing room, ornate or in questionable taste. It should have an appearance that is serious and elegant at the same time.[59]

How it all began: Olivetti's first typewriter, the 'M1', 1911. Designer: Camillo Olivetti.

This philosophy was developed, and was the foundation of a formal design policy, by Camillo's eldest son, Adriano, who began assuming command of the company in the 1930s. Also an engineer, but more intellectual than his father, Adriano was strongly influenced by the Modern Movement theory of the need for a marriage between Man and Machine. He gathered round him a group of likeminded architects, designers and planners; among his reforming actions was the commissioning in 1936 of Italy's first regional planning scheme, two years before he took over as president of the company. It was under Adriano that Olivetti first established a tradition of commissioning top architects to design ultra-modern buildings, ranging from its original Ivrea factory to a celebrated 'anti-industrial' factory complex outside Naples in 1955, and the company's new German and British office buildings in the 1970s.

Among the bevy of architects and designers whom Adriano employed in the mid-1930s was Marcello Nizzoli, an interior designer who quickly became Olivetti's chief resident industrial designer – a position he held until the late 1960s. With an approach to design which he called 'sophisticated simplicity', in stark contrast with the decorative mania which was sweeping the US at the time, Nizzoli went on to integrate his design team deep into the corporate hierarchy. Two of the crowning achievements of his career were the famous Lexikon 80 table-top typewriter in 1948 and the slim and lightweight Lettera 22 portable in 1950. Both count among the greatest commercial successes Olivetti has ever had.

As part of the company's policy of using its top designers to provide 'forward-looking marketing input', as one of today's senior executives puts it, they have worked since Nizzoli's day as consultants, spending about half their time on outside projects – an arrangement which itself helps to give them internal clout within the company. The advantage of securing an outside view also explains their location (in separate studios) in Milan, the main hot-house of Italian industrial design, rather than in the altogether calmer atmosphere of Ivrea, an hour's fast drive away.

The envy of industrial designers all over the world, Olivetti's consultants used to have direct and frequent contact with the company's president. These privileged days are past, thanks partly to the company's growth and the consequent pressures on top management, partly to changes in personalities, and partly to a decentralization of the corporate structure. Some designers consider this a cause for regret,[60] but others welcome it as a maturing of their function from a tender flower in need of top management protection, into a strong, mature plant which has taken its rightful place alongside marketing, development and the other corporate functions, and – though still strongly supported by the top – can more than hold its own.

The consultants continue to be represented near the top of the company by a full-time executive with similar status to the directors of marketing, development engineering and production.

In their shift from classic mechanical products to fully electronic 'ET' typewriters and computer equipment, the twin studios of Sotsass and Bellini lent invaluable assistance to the company's successful conversion from mechanical engineering to the modern

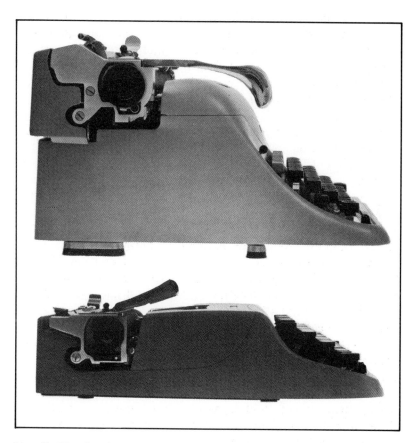

Marcello Nizzoli's classic typewriters. Top: Lexicon 80 desk machine, 1948. Bottom: Lettera 22 portable, 1950.

world of electronics. (Their work for Olivetti grew dramatically during the 1960s, and they formally assumed Nizzoli's mantle upon his death in 1969.)

Apart from the ubiquitous IBM, few traditional office equipment manufacturers have managed to master this daunting revolution. Yet until the transformation of Olivetti's financial fortunes in the years after 1978 under the entrepreneurial leadership of the swash-buckling Carlo de Benedetti, its well-publicised dedication to design was understandably seen by many outside sceptics as an unconvincing luxury. For one thing, not all its stylish products worked very well: a bright red portable typewriter called 'Valentine', which Sotsass designed in the late 1960s, and which was very fashionable for a time, shook annoyingly in use. It had a commensurately short commercial life.

More seriously, the supposed message that 'design pays' was hardly borne out by the company's lacklustre performance. Or was it? Olivetti's director of industrial design argued at the time that the company would have made even less money if it had not been for its high reputation for design. Its past problems were caused by other factors, he insisted, such as its electronics technology not always being as good as it should have been.[61]

This commitment is confirmed by the company's senior engineers and general managers. With advanced electronic technology becoming a virtual commodity to which many companies can gain access, they recognize that product development in the office equipment business has become much more market-orientated than it used to be. In the words of one engineer, 'you cannot bring in the industrial designer at the end of the development process when you're dealing with such questions as how an operator wants to interface with the machine'.

There can be no clearer illustration of this than the experience of George Sowden, a top designer in the Sotsass studio, and one of a dozen consultants working for Sotsass or Bellini (half in each studio); together with another three dozen designers and a minimum of support staff, these 'stars' – who also include Michele De Lucchi – form Olivetti's total industrial design complement (including draughtspeople) of about 50.

One of the growing band of British designers who have left the UK for more fertile work with design-conscious companies in Continental Europe, America and Japan, Sowden first came to public prominence in 1981 as a member of Sotsass' 'Memphis' group of avant-garde designers, whose garish and irreverent products took the international art and design world by storm.

Belying the doubts which business people often have about the ability of 'arty' designers to integrate themselves into the corporate world, Sowden showed an acute awareness of the need for designers to form political alliances with line managers, even in companies where top management has supported the design function as strongly as at Olivetti.

'I still have to form my own relationships with the engineers and persuade them of what I think should be done', he said after Memphis hit the headlines. 'It's not always easy – not all of them have heard of Sotsass, let alone me.' Yet, well before he became an international celebrity, he managed to champion all sorts of design improvements through the corporate hierarchy.

Like his design colleagues, he had spent much of his time out in the field, doing direct market research in the form of talking to bank clerks and computer operators, and watching the way that they work. He was particularly emphatic about the mistake some equipment suppliers and users have made in treating the computer just as a modern typewriter. In fact people use it completely differently.

Detailed observation of clerical reactions to computers in UK client companies such as British Leyland, Barclays Bank and the Abbey National Building Society (one of Europe's largest mortgage banks), led him in the late 1970s to design a swivel device for terminal screens, and more recently a retractable arm. 'Once we'd done that', he recalled, 'all the operators said the same thing – "at last I can push the thing away from me".'

It was through this sort of fieldwork in the mid-1970s that Sowden began to observe the irritation and tiredness which computer operators were suffering because of always having to raise their hands off the desk in order to reach most of the keyboard. This, and early mutterings of complaint from employee representatives, especially in ergonomics-conscious Germany and Scandinavia, prompted him to suggest that Olivetti should radically thin-down the height and profile of its keyboards as soon as possible. But he was too far ahead of the technical standards of the day, and of the thinking of the ergonomics experts within Olivetti's engineering departments. A much slimmer keyboard would also have required a costly redesign.

He made little progress with the idea until 1977, when Olivetti was asked by one of its largest Danish banking customers to design a tailor-made keyboard which would anticipate the standards towards which the country's trade unions were beginning to move. Sowden took the opportunity to press his engineering and marketing colleagues in product development – and their superiors – to transform what was a one-off, custom-built design into the standard keyboard for the new line of 'L1' office computers which Olivetti was in the process of developing.

'A lot of people in the company didn't believe in my design', he recalled – not just for cost reasons but also because it meant a radical change to the design of the keyboard mechanism. At the time, this was still based on Olivetti's mechanical (though very sophisticated) typewriter key mechanisms. The engineers took a lot of persuading that a reduction in thickness by three-quarters (from 120 mm to 30 mm) would not impair the feel and performance of a keyboard.

Launched in 1980, Sowden's keyboard was immediately a world-beater, establishing new ergonomic standards which its competitors were forced to follow – several US companies publicly admitted as much. It may be something of an exaggeration for one Olivetti designer to say that 'if this company hadn't had its keyboard, it wouldn't have sold computers', but the commercial leverage provided by the keyboard was certainly enormous.

A smaller, but equally powerful, example of today's role of industrial design in product development at Olivetti is provided by the way Sowden took the lead, also in the late 1970s, in the development of 'tilt-and-turn' mechanisms for screens on the L1 range of terminals. Another element of the advanced ergonomics which has become such a significant element of Olivetti's success in the cut-throat computer market, it was nevertheless a hard-fought innovation.

There was considerable internal conflict about whether or not the device was necessary, and how much it would cost. With the development engineers and the financial controllers insisting on a very tough cost limit per piece, it might have been dropped or delayed had Sowden not formed an alliance with a particular engineer. Together they refined the design and championed it through the decision-making process.

With the hardware technology of electronics now becoming a virtual commodity, and 'user friendliness' growing ever more weighty in the purchasing decisions of consumers – be they individuals or large corporations – Olivetti's industrial designers may have an unprecedented opportunity to reinforce their influence. But behind this opportunity also lies a threat. An increasing proportion of computer functions are being written in software, rather than hardware, and as a result the 'user friendliness' (or otherwise) of software is becoming a key element in purchasing decisions.

Yet, as in every other computer company in the world, software is the one area on which Olivetti's designers have found it difficult to get their hands. This is hardly surprising, since the one big battle which industrial designers still have to win is with the software writers, a community of exceedingly well-paid specialists who are still considered the 'kings' of many computer companies.

In the past such boffin-like software 'freaks', as they are often called, have frequently been impervious to anything other than technical considerations. Hence the incomprehensibility of much computer software during the 1970s and early 1980s – and even, to some extent, today.

Criticisms of this kind were made, for example, about the operating system and application programs offered with Olivetti's first personal computer, the M20, which was launched early in 1982 under the advertising slogan of 'Brains and Beauty'. To cite just one of the grouses about it, the editing of text was a needlessly complex process.

To some extent such shortcomings resulted from the speed with which the computer and its associated software had to be developed during 1981 – in less than a year, half Olivetti's previous development cycle. Very tight cost controls also played their part.

Together with a marketing decision to keep the machine clearly distinct from the company's L1 range of more sophisticated systems, the tough cost specifications also resulted in the M20 being given a thick, built-in keyboard which was much less satisfactory than Sowden's thin and detachable board for the L1 line.

By mid-1983 senior executives at the company's Advanced Technology Center in Cupertino, California – close to the headquarters of Apple Computer – were conceding some of the criticisms about the M20's software. The machine was replaced in 1984 with the M24, a design which was much more satisfactory in terms of both hardware and software: not only was a detachable, thinner keyboard introduced together with a smaller 'system module' (the box of electronics located beneath the screen), but much of the software was improved, and the operating system made compatible with IBM's. It was promoted with such slogans

Overleaf: *Two generations of Olivetti personal computers. Top: the 'M20', 1981. Bottom: the more advanced 'M24', with separate, thinner keyboard. Designers (of both products): Sotsass, Macchi Cassia, Sowden.*

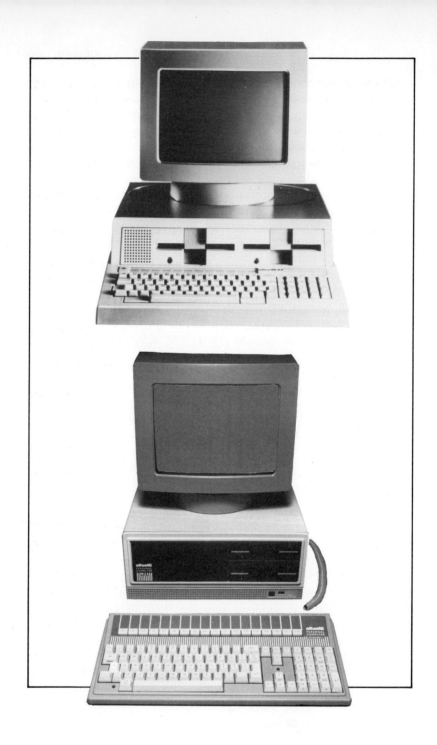

as 'Friendly and Compatible'. (A different version of the M24, using industrial designers from AT & T's 'house' consultancy, Henry Dreyfuss Associates, was used by AT & T as its launch vehicle for entry into the American personal computer market in 1984, and was also sold by Xerox for a time.)

These hardware enhancements, and others which followed later in the 1980s, owed much to Sotsass' studio in Milan. But the software improvements were very much the work of the software experts in Cupertino, where there was still a marked reluctance to allow industrial designers – from far-away Italy or anywhere else – into the software-writing process.

The view both from California and from Olivetti's Ivrea head-quarters since then has been that consumer demand for much greater user friendliness in software has already been making itself felt with increasing force, and has begun to have a dramatic effect on the previously inward-looking software writers. In Olivetti's case this has occurred directly via the example of competitors – notably Apple's Macintosh software – and indirectly via regular and frequent review meetings in which software people are very much involved. The situation has also been helped by the fact that the vast majority of Olivetti applications software is now done not in California, but in Italy.

Despite the lack of official executive enthusiasm about industrial designers getting involved with software, they did in fact gain some influence by the mid-1980s, for example on the writing of videotex software. Now that their ranks have been swollen by some of the new generation of computer-literate design graduates, this influence has started to grow further. Direct pressure from the marketplace is not enough on its own: it may tell a company that its software needs humanizing, but not how to do it – not, that is, unless the company is prepared to rely entirely on copying its competitors, and forever remain one step behind them. A company like Olivetti is not prepared to be a blind follower, which is why it needs the specialist ergonomic assistance of its industrial designers.

Most of Olivetti's designers argue that the growing standardiz-ation of computer technology bodes well, not ill, for their profes-sion. With the marketplace growing increasingly crowded, and meaningful distinction (in Theodore Levitt's sense) therefore at a premium, the industrial design content in computers is set to rise sharply – software included.

Yet the risk remains that, in computer companies which do not enjoy Olivetti's long tradition of powerful industrial designers, influence over software may continue to be denied them. In such cases the industrial designer's role is likely to be confined to making improvements in the comfort of terminals and work stations, and to cladding the hardware in attractive shapes and colours. When average comfort levels reach such a height that there is little further scope for improvement, the designer may be forced back into the role of latterday Detroit stylist, with the task merely of 'putting pretty boxes round things'.

As the imaginative and highly profitable contribution of Olivetti's industrial designers continues to demonstrate, the profession deserves better than that. So do the companies for which it adds so much value, and in whose hands the decision lies.

6 Midwest Pioneer:
John Deere's Company

One day in early 1937 a man wearing a fur coat and straw hat showed up unannounced at Henry Dreyfuss' studio on New York's Madison Avenue, and asked to see him.

Dreyfuss, whose industrial design consultancy had been a remarkable success in its first eight years of life, must have thought his visitor was a drama producer or an advertising copywriter. His name, Elmer McCormick, reinforced the impression. But McCormick declared himself to be the chief engineer at the tractor works of John Deere, a farm equipment company from out in the hick mid-West. His mission, he said, was to persuade Dreyfuss to make the long train journey out to the factory in Waterloo, Iowa, and 'style' Deere's line of tractors.

Dreyfuss was intrigued. He had already worked on a range of products as varied as flyswats, clocks, telephones, typewriters and washing machines, and had just started a major project for the New York Central Railroad on the soon-to-be-famous Twentieth Century Limited. Within a few months he had signed up with Deere and was involved in designing a sleek, streamlined appearance for the company's two main tractors. But his reason for accepting the job was not merely that it gave him the opportunity to extend his considerable styling skills to such unlikely products as farm equipment. He also wanted to apply his credo of improving both form and function as an integral whole (see chapter 2).

The initiative for Deere's espousal of design stemmed not from the top of the company, as at Olivetti, but from the engineers at the Waterloo tractor works. Charles N Stone, Deere's head of manufacturing, said in 1937 that he 'didn't think much of the idea, but

Before and after Dreyfuss. Top: John Deere Model '6' tractor, 1937. Designer unknown. Bottom: John Deere Model 'A' tractor, 1938. Designer: Henry Dreyfuss.

the boys up here want it'.[62] The exact reasons for their enthusiasm are not on record, but seem to have sprung from Dreyfuss'
well publicized, and unusually well-founded, work on streamlining.
Deere's engineers had already discovered the functional value of
streamlined wheel shields which they had added to a line of
tractors launched the year before McCormick's journey to New
York.[63]

From that mission began a remarkable marriage of rugged
engineering and imaginative industrial design which has distin-
guished Deere and Co from most of its competitors ever since.
Its far-reaching use of designers from Dreyfuss' consultancy,
especially since the mid-1950s, has not only given it a glamorous
corporate image, but has improved the performance of its broad
range of products: Deere is highly reputed among the farming
community for being a leader in integrated attachments for tractors – comfortable seating, easy-to-use controls, high standards of
cab insulation, safety features of various types, and other aspects
of the 'human engineering' which Dreyfuss pioneered.

Deere's rise to become America's largest maker of agricultural
equipment, and to dominate the two prime product segments
(tractors and harvesters) has been rightly ascribed to several
factors: technical innovation (initiated by the company itself, as
well as in response to farmers' demands for new features); high-
quality manufacture; the building of a strong dealer network
and, underlying everything, the will to invest long-term. But the
company's top management is also convinced that industrial
design has played a key part.

This is true not only of the long halcyon years of American
postwar agriculture, when the cost of equipment was no object to
the featherbedded farmer, but also of the tough 1980s, when it
suddenly became difficult to make a sale, especially one which
earned a profit. As the entire US farm equipment industry plunged
into loss, Deere's ability to differentiate its products through
design – both form and function – became even more valuable
than before. And in Europe, where fragmented markets and a
patchy dealership network stood in Deere's way until the 1980s,
design has played a crucial part in the company's belated rise
to prominence. Today, Deere's characteristically dark green and
yellow machines have become a common sight in European fields.

To a surprising extent for something as functional as a tractor or a combine, Deere has found over and over again that appearance does matter. Just like ordinary folk with a new car, it seems that farmers are prepared to splash out on an attractive new machine before its utility and reliability are really proven. In 1970 Deere's share of the US harvester market was given an instant and massive boost by the launch of a new product line. In 1983, when the company was test marketing its top-of-the-range Titan combine in France, it found that farmers in the fertile Loire valley were particularly impressed by the rugged looks of the machine.

Back at Deere's headquarters in the town of Moline, not far from the banks of the great Mississippi, the product engineers will tell you of the great pride of ownership that exists among farmers: 'a guy will take you out to his barn just to show you the carpeting in his cab'. And they will refer to a research study carried out at a local university which found that appearance does indeed rank highly in farmers' purchasing decisions.

To the outsider, the impression that this may be merely a matter of styling, borrowed by Deere from nearby Detroit, is strengthened by the way that the company's executives describe the consultants from Henry Dreyfuss Associates – up to four or five are involved at any one time – as 'stylists'. But this is misleading. In the main, it seems to be an unconscious way of distinguishing their role from that of the company's own design engineers. Sometimes referred to simply as designers, the latter have traditionally constituted the most powerful functional group within the company. (Their possession of more corporate 'clout' than production, marketing or finance can be traced back to the company's foundation in 1837 by John Deere, a blacksmith who designed the world's first successful steel plough.)

In fact, Deere and Co is imbued with the Dreyfuss philosophy that form and function are inextricably intertwined: that 'if a product is streamlined for appearance, it's streamlined for function, too', to quote Russ Sutherland, who as Deere's director of product engineering husbanded the link with the Dreyfuss consultants for many years, before being promoted to the vice-presidency of engineering as his final job before retirement.

Certainly, the consultants' work has for many years been far more weighty than mere styling. They may not initiate entire product concepts, as do designers in some less engineering-intensive companies, but they have had a considerable impact on Deere's technical decisions. Within what is still a highly decentralized company, they also play an important role as cross-pollinators of ideas between different units.

The Dreyfuss influence on Deere's engineers was felt as early as 1938 when the company was persuaded to change the steering mechanism for one of its tractors so that it no longer protruded from the hood (bonnet); the arrangement, suggested by Henry Dreyfuss himself, was also cheaper to make.[64] There has been a steady stream of such trade-offs between engineering and industrial design ever since, involving not only finished products (such as the system for oiling a harvester) but also components (such as the pattern of liquid flow within a hydraulic pump).

The hand of Dreyfuss and his team was strengthened in the 1950s by two developments. One was the gradual emergence of ergonomic requirements as an important factor in product design. The other was the arrival in 1955 of a new chairman and chief executive, William Hewitt, who had an unusually strong interest in art, architecture and design.

Though Deere's initial decision to make use of industrial designers had originated from a relatively low level in the organization, top management's commitment and evangelization later became as much a part of the story as it has at other design-minded companies.

The son-in-law of the previous president (himself a great grandson of the founder), Hewitt had spent several years as an advertising copywriter before joining the company. He went on to stay at the helm for 27 years, and exercised a dominant personal influence until he retired in 1982 to become US ambassador in Jamaica.

Until Hewitt's arrival the company's commitment to industrial design had been rather patchy outside Dreyfuss's main 'base', the Waterloo tractor works. As a junior design engineer, Russ Sutherland remembers using a Dreyfuss consultant to help design a siderake in 1951; and he can recall the company's product planning chief introducing Dreyfuss to various unit heads at about the same time. But at that stage it was generally left to individual design engineers to decide whether, and how, they would use the Dreyfuss people. To the extent that there was a corporate

Eero Saarinen's award-winning 1964 headquarters building for Deere & Co, Moline, Illinois. Its significance to the company extended far beyond the quality of its design.

commitment of any sort, it emanated mainly from middle management, not from the top.

With Hewitt, all that began to change. He applied a strong sense of the visual to all the company's activities. Within two years of moving into the dingy old corporate headquarters behind the railroad tracks in downtown industrial Moline, he had commissioned Eero Saarinen, the world-famous Finnish-born architect, to design a new complex of buildings to be set in the green, rolling hills outside the town.

Completed in 1964, it is rightly considered to be one of the most spectacular yet functional headquarters of any multinational company. Amid beautifully landscaped lakes and gardens, with the inevitable Henry Moore sculpture, it is in every sense stunning, with soaring rust-coloured steel beams and sunscreen louvres enclosing acres of glass. Inside, every detail is carefully designed. Its bright, long corridors are closely lined with tapestries, sculptures and paintings, many of them originals, and the classless cafeteria is sited inside an atrium-like structure complete with full-size indoor trees. Saarinen was blazing a trail which others took 20 years to follow, and which in the meantime won him countless architectural awards.

To the company, the significance of the buildings extended far beyond the winning of prizes. In his definitive history of the company, Wayne G Broehl, Jr provides a classic justification of the case for design-mindedness in environments, as well as products.[65] When Hewitt put Saarinen's costly scheme to the board, says Broehl, he was

asking the board to enlarge its concept of the company – to see itself as a major, nationally important entity, on its way to becoming the pre-eminent firm in the industry. In effect, a statement was to be made by the company, one which Hewitt hoped would reverberate through the organization at all levels and out among the dealers, the customers and the general public . . . It is probably not excessive to say that at this meeting Hewitt turned this insular, Midwestern farm implement company in a new direction, pointing it toward a new role as one of the world's best-known multinational corporations.

Hewitt's brief to the architect made it clear that the complex should be in harmony with the way the company saw itself, and with the function of its products. Inasmuch as the farmer 'wants and needs the most efficient and durable tractors and implements', he wrote, 'we also want and need a headquarters building that will utilize the newest and best architectural and engineering concepts'. Therefore 'the several buildings should be thoroughly modern in concept but should not give the effect of being especially sophisticated or glossy. Instead, they should be more "down-to-earth" and rugged'.[66]

Hewitt also paid close attention to Deere's documentation, right down to the quality of the graphics used by managers in their internal presentations. But it was through the company's growing line of products that his commitment to design became most evident.

It is not surprising that someone with such an interest in design, and dedication to it, should have become a personal friend of his chief design consultant. But it was the commitment, rather than the friendship, which caused Hewitt to encourage the idea that Dreyfuss and his staff should be called into product development projects right at the start, where their influence could be greatest. It also frequently prompted him and his senior lieutenants to seek direct advice from them, either in person or on the telephone.

In spite of this strong top-down support for the Dreyfuss work, all was not plain sailing. A few factories refused for a time to use the designers, on the grounds that 'if our engineers are good enough, we won't need them'. As late as the 1960s the chief engineer at one works near the Moline headquarters was resisting along these lines. It took some time for the industrial designers to overcome such resistance, partly by biding their time until more receptive managers took control at works level, and partly by rolling up their sleeves and working extra hard alongside the engineers in order to prove themselves.

By the mid-1960s the value of the Dreyfuss contribution had been made manifestly plain for all to see. For in 1960 Deere successfully launched what it called its 'New Generation of Power' – the '10' series of tractors. The first major product line to be developed entirely after Hewitt's rise to power, the tractors not only contained countless engineering improvements on their predecessors and on the competition, they also benefited from an unprecedented degree of attention on the part of the Dreyfuss designers.

The closeness of the collaboration is described in detail in a highly illuminating article by William F H Purcell, an early colleague of Dreyfuss, and for many years the consultancy's chief link with Deere.[67] The Dreyfuss team – consisting of Henry Dreyfuss himself, Purcell and a design associate with eight years background in the design of farm machinery – was brought into the development process from the very start, alongside Deere's top management and their chief researchers and engineers.

At the very first meeting Dreyfuss suggested that the fuel tank should be mounted in a position which did not protrude awkwardly through the hood, as with existing models: it was decided to try to place it in front of the radiator (a solution which was eventually adopted).

More significant in functional terms was the application of 'human engineering' (ergonomics) to every aspect of the design, especially controls and seating. Referring to the famous Dreyfuss drawings of the 'average' American couple, Purcell commented that 'Joe has been consistently cared for. He can mount and

Two generations of Dreyfuss-influenced tractors. Top: '10' Series, 1960. Bottom: '50' Series, 1972.

dismount with ease, stand or sit with safety and comfort, and has good visibility'. (Presumably the same went for Josephine, though on this occasion she did not rate a mention.)

The biggest breakthrough of all was the new and highly adjustable 'posture seat', designed by Dreyfuss consultants and

Deere engineers with the help of a leading medical expert whom Dreyfuss introduced to Deere. It was so comfortable that farmers sent fan mail praising it.

The industrial designers played an equally influential part in developing the next line of tractors, 'Generation II', which was launched in 1972. Here the prime advance was the 'Sound-Gard' cab. Designed as an integral part of the vehicle, rather than as a cab just to be tacked on afterwards, it had high standards of vision, sound insulation and safety. Similar ergonomic improvements were applied to the line of 'Titan' combine harvesters launched in the US in 1979, together with an altogether 'cleaner' and smoother look to the machine's sheet metal panelling.

Among additional ergonomic features initiated by the Dreyfuss consultants were the introduction of a 'grease bank' within the machine to replace the old system of greasing individual parts each day. Purcell spurred this fundamental improvement in design-for-servicing by arranging a greasing demonstration by a man in white overalls – the results were predictably messy. Another functional improvement urged by the Dreyfuss designers, on the grounds that a longer rear 'hood' would look better, was a lengthening of the straw walkers within the machine.

On the other hand, the industrial designers failed, on cost grounds, to justify another functional improvement which they had advocated: the addition of a small hydraulic cylinder to provide enough power to fold the harvester's access ladder: this still has to be done by hand.

On the development of the 1979 harvester, as with the 1972 tractor, the Dreyfuss team was called in at the very start of the process, to help create the thick specification manual and to work hand-in-glove with the engineers from that moment on. This was a major departure from the procedure adopted for the previous generation of harvesters (launched in 1970), where a good 18 months of crucial decision making had elapsed before they were called in.

Yet even then they had exercised a key influence over the machine's configuration: the positioning of the engine at the front, rather than at the rear as in most competitive US harvesters (and even today in Deere's own European-made combines, which are less sophisticated than its US products). A change had been proposed on technical grounds, in order to lower the machine's centre of gravity, but the marketing department was nervous that it would prove too radical a change for the customer. The decision was finally upheld after work by the Dreyfuss consultants had demonstrated that the prototype with a front-mounted engine looked more attractive.

Involvement of the Dreyfuss consultants at the very start of the development process, in close alliance with Deere's product engineers, has given them unusually strong influence. But it has not quite turned them into the second most powerful function in the corporate hierarchy. For the product engineers no longer rule the company as they once did, and product development is now a collaborative process which also involves marketing, production engineering and finance. So the political process which the consultants must master has grown more complex.

It is further complicated by the changes which have occurred at the top of the company since Hewitt's retirement in 1982. Over this period the degree of direct top-level involvement has declined, as at Olivetti, and most of the contact is back at unit and factory level. Whether or not this bodes ill for the influence of industrial design at Deere remains to be seen. For most of the 1980s the company was grappling with the vicious pincer effects of rising costs and falling demand, and the attentions of Hewitt's successor, Robert Hanson, were inevitably directed elsewhere.

Even at the elevated level of executive vice-president it is conceded that Deere's tough cost-cutting drive probably put pressure on its traditional dedication to stylish design in every aspect of its products; a number of senior engineers, especially the younger breed, indeed questioned the need for quite such a degree of emphasis on design details such as rounded, rather than square, corners. Since these require extra milling, they can carry a noticeable cost penalty.

But the influence of industrial design at Deere seems safeguarded to some extent by several factors, most notably the widespread recognition that the Dreyfuss consultants have always tended to emphasize that designs should be simple, not complex. Examples abound of their proposals helping to cut the cost of a product, rather than increase it. One is of a disc harrow, where a vital $20 was shaved off the manufacturing cost by bending one piece of tubing to make the frame, rather than welding four pieces together.

A degree of protection is also provided by Deere's traditional system of not charging individual units or factories for the use of Dreyfuss consultants, but billing headquarters direct. Both sides agree that this is one of the keys to the success of the relationship through good times and bad.

It is debatable whether more is lost or gained by the external status of the Dreyfuss team, and their location as far away as New York, several hours' flying time from the Illinois/Iowa border, where most of the Deere facilities are located. Some Deere engineers consider that less distance would bring the consultants more influence, especially now that the company's development time scales are having to be compressed in order to cut costs and stay ahead of the competition. But in the US it is difficult to achieve the close proximity between consultant and client which is considered necessary in Europe, and practised to such good effect in the case of Olivetti.

On the other hand the disadvantages of distance are at least partly outweighed for Henry Dreyfuss Associates by the added perspective and influence which their New York base provides. In the words of one of Deere's vice-presidents 'they not only provide a detached, objective view, but also a depth of vision that's not hindered by existing ideas and investments. When they show up [and they do, often], our people pay attention.'[68] Coming from a company with such all-round management excellence, that is high praise indeed.

7 Harnessing Flair:
Sony

Sony has been a watchword for good design for at least a quarter of a century. So it may seem churlish to classify this most innovative of Japanese companies as a design convert, rather than as a pioneer.

The company's legendary co-founder and chairman, Akio Morita, has always possessed an instinctive understanding of the power of design to enhance Sony's corporate image, as well as its range of audio equipment, televisions and video products. He hired his first industrial designer in 1954, when his company was only eight years old, and by the early 1960s, when it first began to develop into an international corporation with a household name, it employed 17 designers – more than many companies several times its size. And via Norio Ohga, who was appointed president in 1982 after 23 years of steadily increasing influence throughout the company, it had for many years a near-Germanic commitment to functionalism (Ohga spent several formative years in the 1950s as a student in Germany).

Yet it was only in the late 1970s that Sony really began to use industrial design as an initiator and integrator of product development. During its crucial growth decade of 1968–77 its designers had been based out in the audio, TV and professional equipment factories, where they were under the thumb of engineers, salesmen and short-term-minded product planners.

As late as 1982, at celebrations in London to mark the opening of the first international exhibition of Sony's design, Morita was still tending to slip into a narrow definition of design as 'very important: if a machine looks ugly, it won't sell'. A similar formulation, on the same occasion, was that 'the appearance of the product has to be good – that's why we're spending so much effort on industrial design'.[69]

Such lapses, at odds with the growing reality of Morita's use of design since he had centralized it under a powerful manager in 1978, can only be explained by his characteristic tendency to stress the glamorous exterior of his company, rather than its complex inner workings. He is, after all, a showman par excellence who knows better than anyone how to attract publicity.

In fact, Morita had been personally instrumental, along with Ohga, in steadily raising the status and influence of industrial design. By the time of the London exhibition[70] it had begun to act not only as a rich fountain of new product concepts on its own account, but also as an umbrella for the development of new ideas brought in by staff from various product divisions. And it was starting to help coordinate product development right across the company, a task made increasingly vital by Sony's growing size and complexity, and by the pressing need to integrate the previously separate technologies of audio, video, computers and communications.

This additional role was formally recognized in early 1985 when Sony's design chief, Yasuo Kuroki, added 'Director of Consumer System Products' to his title, and a bevy of senior engineers and planners to his staff. An extension of this magnitude in the formal role of design is virtually unparalleled, both in Japanese industry and in the West. Sony has certainly become a latterday design pioneer.

Its radical upgrading of design was prompted by immediate pressures as well as long-term considerations. As the 1970s drew to a close, the sales growth of many of Sony's staple products – including Trinitron TVs and much of its audio equipment – slackened, leaving the company increasingly reliant on the profits of its Betamax video cassette recorders (VCRs). But though the Betamax system had enjoyed a head start when it was launched in 1975, it began to be overtaken in most markets by JVC-Matsushita's VHS system, which has sold round the world under a plethora of brand names.

Sony's extraordinarily successful range of Walkman personal cassette players, launched in 1979, provided a degree of compensation, as did several other new products. But Morita had already

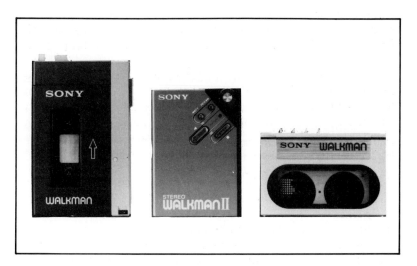

The 'Walkman' phenomenon in three of its many guises: the original product, launched in 1979 (left) has been miniaturized and revamped several times, giving Sony a wide range of models.

begun to recognize the need to expand the company's scope in consumer electronics, as well as to break into computer and office products. He had also realized the necessity of augmenting Sony's traditional sources of innovation: the technology-driven work of its development laboratories and the fabled marketing flair of himself and his close advisers.

In particular, Morita was beginning to be aware of the need for Sony to develop a greater depth of understanding of changing social attitudes and behaviour, in order to develop more of what the company's executives call 'software thinking'. Their use of the term has little to do with the narrow world of computer software. Instead, borrowing their terminology from J K Galbraith (one of the many Western influences on Sony), they explained that 'up to now Japan has only produced *products*, paying little attention to considerations of culture and life style. We must change'.[71]

In a company with particularly little faith in traditional methods of market research, the ability to anticipate and exploit changing social behaviour is a crucial complement to technology-driven innovation. Sony's 'feel' for how it can mould the consumer has always been a prime factor in its commercial success, from the transistor radio in the 1950s, through small black and white TVs in the 1960s (see chapter 3), to VCRs and the Walkman in the 1970s.

This knack, in which market research can at best play a subsidiary role, will take on even greater competitive significance in the future. But within Sony it has always rested very much with individuals, and Morita recognized the need to prepare for the day when both he and his co-founder Masaru Ibuka (the engineering foil to Morita's marketing genius) had retired from the scene. An institutionalized element had to be added to the company's creation of its famous flair.

It was to underline top management's changing view of design that the English language title of the new department was changed, within six months of its creation in June 1978, from 'Design Division'. The utterly confusing replacement, 'PP Center', did not, as many outsiders assumed, refer merely to 'product planning' – this was a separate, short-horizon function which had existed within Sony for many years, as a back-up to production and sales. Instead 'PP' was left intentionally vague, to mean anything from product presentation to proposal and promotion – or, indeed, advanced product planning.

Under the reorganization, the majority of designers retained their location in the divisions, but it was made very clear that they now reported direct to Yasuo Kuroki, who had been promoted from heading Sony's domestic advertising department to the top of PP Center. The overall complement of the design staff was increased by more than half to over 100; by 1984, when a further degree of centralization began to take place, it had risen to 130, of whom 70 were design professionals. Two dozen were working at Sony's Tokyo headquarters, about 20 in the audio and video divisions, a dozen in TV and about 20 in 'professional' products.

The most immediate result of the change in structure was much better cross-fertilization between designers in different parts of the company. Previously they had seldom held meetings, and there was very little mutual assistance and criticism. Kuroki quickly initiated an intensely interactive process of three tiers of formal weekly meetings involving the entire staff, plus all sorts of informal communication channels. In common with many European companies, but in contrast with most large US enterprises, Sony benefits from having most of its factories and laboratories within reasonably close proximity of its headquarters.

One of the first innovations from Sony's 'PP Center': the 'Profeel' Component
TV System, 1980.

From the moment his department was formed, Kuroki had a
remarkable influence on top management. Not only did he enjoy
direct and open access to Morita, Ibuka and Ohga, but he became

the only department head to receive visits every month or two from a delegation consisting of the company's top half-dozen executives. With assiduous regularity, they have trooped down to the designers' offices to inspect the latest ideas. This 'creative report', as it is officially known, generally lasts an hour and a half, and focuses on about half-a-dozen products each time.

In spite of this strong backing, frequently reinforced by documents carrying chairman Morita's own signature, it would have been surprising, given the realities of managerial life, if Kuroki and his staff had enjoyed a completely smooth ride with their colleagues in marketing, engineering and production. But they certainly developed the clout to discuss, and if necessary, argue with them. In most other Japanese consumer electronics companies, design still tends to be dominated by marketing, and especially by development and production. At Sony, the designers are not only expected to master engineering details, but also marketing techniques, right down to making detailed proposals for retail pricing.

For Sony's designers it is not relations with marketing which have been the main source of internal difficulty, but those with production and sales. This classic tension between innovation and the status quo beset PP Center's very first product concept, the 'Profeel' component TV.

A modular TV system in which high quality colour monitor, receiver, tuner and speakers were split up from each other, the Profeel had Morita's personal blessing right from the start (indeed, he is sometimes attributed with joint parenthood of the concept). But production refused to lay down capacity, on the grounds that the sales staff doubted whether the product would sell.

In an illuminating display of Japanese 'Nemawashi' (which literally means 'root-binding', or what Westerners call 'consensus building'), Kuroki then had to set to work convincing the doubters that consumers would indeed want such a futuristic product even in the absence of many of the services which it would eventually carry: satellite broadcasting, cable TV, stereo sound, interactive home computers and so on.

The Profeel could have been on sale by mid-1979, but the opposition held it back until the following February. Yet within three years it was accounting for nearly 5 per cent of Sony's world-wide TV sales, and over 10 per cent in Japan. In the depressed TV market of the time, that was a most useful extra volume of

Another 'PP Center' initiative: the 'Watchman' flat-screen pocket TV.
Left: first-generation, pre-launch model, 1982. Right: miniaturized second
generation, 1983.

business. Moreover, Sony's traditional policy of being first into the
market with an innovative product again enabled it to charge more
than its competitors, who quickly rushed copies into the shops.

PP Center's next project, a flat screen pocket TV which was
eventually christened 'Watchman', after its Walkman cousin, was
originally conceived as a highly specialized product, directed
especially at professional use, including TV programme makers,
both in the studio and on outside broadcasts. Encouraged by Morita,
the designers raised their sights to various other specialized market
segments, including people watching sports events. But everyone,
including the designers themselves, saw it as just a gadget, an
attitude which initially discouraged the company from investing
much in production equipment or the streamlining and cost
reduction of the set's circuitry and components. This allowed
competitors like Britain's Clive Sinclair to describe the set as
'jerry-built'.

When the Watchman was launched in early 1982, after a development cycle of little over 18 months, its success astonished even the most optimistic of the designers. Its initial production run of 2,000 units per month had to be more than doubled by the end of the year, by which time the company had made the necessary investments to redesign the product for reduced battery consumption, better quality and reduced cost. The greater degree of integration of its circuitry made it much smaller and lighter, and altogether more comfortable to carry and use. Launched in the spring of 1983 with a production level of 20,000 units a month, 'Watchman II' was also more successful than expected, and output had to be raised still further.

The interval of a year before demand could be met gave Sony's Japanese competitors valuable time to break into the market. Sony was fortunate that its chief rival, Sinclair, was held up for another 18 months by production problems. In the meantime Sony's readiness to launch such a rapidly developed product had given it a precious start down the 'experience curve' of low cost mass manufacture. And, as so often in Sony's history, the company had been able to create an innovative aura for a new product, and with it the ability to charge a premium price.

The designers continued to initiate a stream of product concepts, ranging from a small security camera to a portable TV video projector. They also took the lead in designing a smaller, second-generation 8 mm video camera–recorder, the mini 'Video-8', which was launched in the summer of 1985.

Examples of design-led product concepts also extend to Sony's designers based outside Japan. For instance, in the summer of 1986 a brainstorming session in San Francisco between a group of US-based marketing and sales people came up with an idea for a product for children. The concept, with one person's suggested title 'My First Sony', was immediately snapped-up by the head of Sony's New Jersey-based North American design team. After his team had developed it into a range of four products – a Walkman, a cassette recorder with microphone, a walkie-talkie headset, and a radio cassette recorder – he took it to Tokyo with a rushed set of mock-ups, packaging and logo, as well as business plans which included estimates of sales volume and projected profits.

Winning support from headquarters was not easy. Engineering and production staff were sceptical about the product's potential – though they liked the designers' innovation of a window showing the mechanism – while one top executive disliked the proposed name, complaining that Sony ought not to give the market the impression that it was a toy manufacturer.

But, as with the Profeel, the designers argued the project through, and it was finally given the go-ahead in exactly the form they had first proposed. The four products were developed in well under a year, and went into production in good time for the Christmas market. They easily reached their projected annual sales rate of 400,000 units in the US that first season, and by early 1990, after just three Christmas selling seasons, had done even better, reaching 1.6m unit sales in the US alone, and a further 500,000 in Europe and Japan.

Another type of design role at Sony, the championing and nurturing of product concepts 'imported' by staff from the various product divisions, is best exemplified by the Betamovie. The world's first portable home video camera–recorder (Camcorder), it was the brainchild of a pair of engineers in the video division. Their own department was reluctant to develop it, being heavily committed to other projects, and aware that a different division (professional products) was working on the development of a similar, though more elaborate, concept.

But Kuroki seized on the project with open arms, and proceeded to play a sponsorship and nurturing role rather like that of the famed corporate new ventures department at 3M. Acutely conscious of the growing consumer preference for compactness and convenience, and of the ripeness of the camera market for such an innovation, he funded the development of a working model which was presented to top management with his enthusiastic support. Approval was quickly obtained, and a final prototype developed within PP Center. With a fanfare of publicity, the Betamovie went on sale in the summer of 1983, and was an immediate smash hit, sparking off a competitive race to develop the next generation of even more compact camera–recorders.

The examples so far cited demonstrate three roles of industrial design at Sony: its regular work on every new product, from compact discs to the latest Walkman derivative; its initiation of product concepts; and its championing of concepts brought in from elsewhere in the company. Allied to all three is a fourth: coordination of the work of other functions, divisions or departments to an

Sponsored by 'PP Center'. Top: Sony's 'Betamovie' video camera–recorder, half-inch format, 1983. Bottom: ultra-compact mini 'Video-8' camera–recorder, 8 mm format, 1985.

extent which sometimes amounts to what many other companies call 'product management' or 'programme management'. (This sort of role is most commonly carried out by a separate managerial function – very rarely by design.)

Just as PP Center's work on the Profeel had entailed considerable coordination of different divisions, so the later stages of the Betamovie project involved the designers in liaison between various units inside the video division. With the convergence of different technologies proceeding apace, more and more projects began to be of this nature. One of Kuroki's tasks in 1984, at the special request of the head of the professional equipment division, was to coordinate the work of several product planning and development groups on a new computer terminal system.

The convergence of technologies, and the consequent need for individual products to become increasingly integrated with each other – or at least capable of integration if the consumer so desired – lay behind the decision in 1984 to give PP Center a greater degree of centralization, and to break down its vertical, product-based structure.

Explaining the move at the time, senior executives in PP Center justified it on the grounds that Sony needed to adopt a much more standardized approach to design. There should still be plenty of room for variety, but it should be planned, rather than haphazard. Audio, video and communications products would not only have to be designed to work well together, but to look good together in the home or office.

The first step in this upheaval was to move almost all the designers to Sony's headquarters, in the Tokyo suburb of Shinagawa. This created some concern in the divisional factories, especially on the part of engineers who had come to rely on close personal cooperation with the designers. But it was persuasively argued that the drawbacks of such an 'arm's length' relationship were outweighed by the advantages. A designer who is located within a plant can come to understand the engineer's point of view almost *too* well: he tends to be dragged into the daily problems of production. One manager described the change thus: 'The designer needs to see from a distance. He can't be creative if he's too involved in the nitty gritty of operations.' [72]

The second stage of the reorganization was the move to a horizontal structure, with much more collaboration between different

specialist designers. This became part of a much broader reorganization and upgrading of the role of industrial design. In order to
reinforce the ability of Kuroki and his senior staff to stimulate
and coordinate development projects which spanned different
divisions, top management decided with effect from early 1985 to
create an entirely new organizational grouping, Consumer System
Products and Design.

Kuroki was made its director, under one of Sony's several
managing directors. The new group not only included the old PP
Center, renamed Design Center and reorganized into just two
divisions, along horizontal lines; it also embraced new teams of
development engineers and marketing/business planners, together
with a new function long advocated by Kuroki, 'merchandisers'.
Their function was to help speed individual projects right through
from initial concept to design, development, production, launch
and sales promotion. Even Japanese companies can suffer from the
inevitable interdepartmental barriers that bedevil the development
process in the West. Merchandisers are one of various devices that
can be used to overcome such hurdles.

In mid-1988 a further reorganization took place, in which
engineering and coordination were officially split off into a
separate group and the design department's main role was
refocused under the new title of Corporate Image Structuring
Group into three sections: design, new business development and
advertising/promotion.

At first sight this seemed to imply a reining-back of design's
influence at Sony, into a much narrower role than it had enjoyed
for the previous decade. Nothing could have been further from the
truth. Not only do official organization structures mean little at
Sony, but Yasuo Kuroki was promoted as part of the change onto
the company's main executive board, making Sony only the
second Japanese company to have an industrial design director at
that level (the first was Sharp).

Moreover, the 1988 move was very much Kuroki's own
idea – he had been talking about it for well over a year beforehand. Furthermore, the design engineering and system coordination departments were moved only just through the wall from
design, so that they were still barely a step away and collaboration
came naturally. Rather than causing any loss of design influence,
the main purpose of the department's changed structure and title

was to signal a particular new direction for a year or two. This continued the shifting pattern of previous years: the invention of the intentionally vague PP Center title in 1978, the move to Consumer System Products and Design in 1985, and now a title focusing attention on corporate image.

The thinking behind the 1988 title was essentially this: Sony had been striving for some years to diversify from consumer products into professional ones as well, and was starting to succeed. At the same time it had also begun moving in an opposite direction, into consumer 'software': first with CBS Records and then, in 1989, with Columbia Pictures. As Sony's top design managers said at the time, this required a carefully controlled shift of corporate image.

By 1990 it had already become a fair bet that, once that shift was deemed to have been completed effectively, the design department would be restructured – and retitled – again. Frequent organizational changes are very much part of the Sony culture – as is flexibility in the way those organizations actually operate.

For many American and European managers, Sony's way of managing design and development is 'bewilderingly chaotic', in the words of one of them. They much prefer the more ordered-looking sort of approach represented in this book, for instance, by Philips (chapter 9). But, as Tom Peters rightly argues (after that influential strategy professor James Brian Quinn, the long-standing luminary of Dartmouth College, New Hampshire), 'controlled chaos' is almost certainly the right management style of the 1990s.[73] Peters sees it as a synonym for corporate creativity, innovation, and competitiveness.

So does Sony. That industrial design should achieve such a position of power within this remarkable company bears convincing witness to the designer's ability to combine imagination with skill, drive and managerial discipline.

8 Breaking the Detroit Mould:
Ford

Ford's cars used to look characterless and utterly nondescript. If they were distinguishable at all from other makes, it was only for their anonymity.

Today it is a very different story. In both the US and Europe, its sleek range of aerodynamic vehicles stands out a mile from the crowd. Speeding along an American freeway or a German autobahn, they really do live up to the extravagant claims made by some of Ford's designers, of 'moving sculpture'.

Behind this change in appearance lies Ford's transformation from design dullard into leader. Starting in Europe in the late 1970s, it completely broke away from its traditional strategy of providing worthy but boring products on a narrow sales platform of 'value for money'. And in 1980 the parent company in Dearborn, Michigan, abandoned its standard approach of plodding along steadily in the wake of General Motors. Instead, the entire Ford organization seized on a nascent European trend towards aerodynamics to leapfrog its competitors, not only in Europe but around the world. In one courageous and risky bound it scrapped the time-honoured Detroit tradition of evolution at a snail's pace, and catapulted itself forward by a generation. As one of its top executives said at the time, 'we clearly identified the need to re-establish our image in the marketplace. You can't do that if you're timid.'

Despite tricky problems in a few European markets, Ford's adventurousness was quickly rewarded. In Europe it fought its way to market leadership for the first time, and in the US its 'jellybeans', as the popular press dubbed its new range of aero-dynamic cars, quickly restored several precious points of market share that it had lost to GM and the Japanese. During the rest of the decade its US market share soared to over 22 per cent, from 17 per cent back in the early 1980s. That was the overall effect. In the particular market segment at which the Taurus was pitched, the 'mid-size' category, the share taken by the Ford brand leapt from 9.9 per cent in 1983–5 to 21.4 per cent in 1987–9.

By 1990 its new range of cars had put it in a much healthier position than GM: despite its much smaller size, it actually earned more profits for several years running in the late 1980s. As a result, although it continued to have difficulty matching the fast product development cycles of its Japanese rivals, it went into the 1990s far better equipped than GM to cope with the worryingly softening market and intensifying competition with which the new decade began.

As was made clear in chapter 1, the revolution in Ford's design was far more than styling. Its new image not only reflected a carefully planned shift away from its traditional marketing strategy (in the proper sense of the term), but also a policy of integrating form and function, as well as an ambitious integration between the company's industrial designers, engineers and product planners. The whole story is an object lesson in the management of product development, and the changes that are necessary in executive attitudes, corporate structures and market research procedures, if a company really wants to use adventurous design to create products of meaningful distinction from the competition.

Europe Shows the Way

The shift in the company's competitive strategy began in the mid-1970s, in its main market stronghold of Western Europe. Faced with an overwhelming Japanese challenge for its traditional platform of cheap and cheerful cars, Ford of Europe's top management in Britain and Germany decided

to get Ford cars out there that people desperately want, rather than cars they will buy because they are the lowest priced on the market. You can't do that any more because the Japanese have taken that part of the market away from us.[74]

The words are those of Robert Lutz, the former BMW executive who ran Ford of Europe between 1977 and 1982 (first as president and then as chairman), and again for a short period after 1984 before moving to Chrysler. In a frank interview in 1982 he admitted that

the Japanese have taken over the no-nonsense, no-frills, high-value for money, reliable transportation part of the market. My goal is to be a mass producer of the type of cars BMW and Mercedes have a reputation for making. We are moving up in technology and credibility so we get the same price elasticity as they have.[75]

Lutz might have added Volkswagen–Audi to the list of companies Ford was trying to emulate. But that would have been too near the knuckle, since it was precisely the upper end of the VW range, and the entire Audi line, at which Ford had decided to pitch itself. This quest for quality was the first plank in its new European strategy.

The second was aerodynamics. By the late 1970s a number of Italian design consultancies had begun to display prototype 'ideas cars' at the leading annual motor shows, fusing together a growing obsession with aerodynamics, the mushrooming requirements of safety legislation, and unusually skilful 'packaging' – the squeezing of as much interior space as possible into tightly restricted exterior dimensions. The Italian prototypes went well beyond the achievements of Citroën and Rover, which had been making aerodynamic-looking cars for some time.

Amid signs that Audi and a number of other competitors were beginning to recognize the commercial advantage to be gained by offering remarkable improvements in fuel efficiency, together with extra passenger comfort and a futuristic new appearance, the then head of Ford Europe, 'Red' Poling, and his new vice-president of design, Uwe Bahnsen, decided to apply some of this thinking to the new Escort model which was already under development. ('Small-to-medium' in European size terms, it is classed as a 'subcompact' in the US.)

But winning the approval of Ford of Europe's all-important product planning and design committee was far from straightforward. Strange as it may seem in retrospect, the committee's heavily conservative membership of product planners, engineers and salesmen/marketers was initially sceptical of the notion that

The first manifestation of Ford's new marketing and design strategy: the 'old' 1970s 'Escort' was replaced in 1980 by a completely new vehicle bearing the same name – and with a controversial 'notchback'.

aerodynamics would produce fuel savings. Still all too aware of Detroit's disastrous experiments with aerodynamic styling in the 1930s (see chapter 2), they were nervous of the 'notchback' which Poling and Bahnsen proposed to give the Escort, in preference to the hatchback shape which VW and Renault had pioneered, and which was then becoming standard in cars of the Escort's class.

By undertaking additional market research, and making more intelligent use of the results than in the past, it was eventually possible to persuade them that its unusual silhouette would not deter consumers from buying the car. And so it proved. Launched in 1980, the Escort was a stunning success in almost every European market.

For Bahnsen's team of designers, it was the big breakthrough: they had broken the ice of established traditions and convinced the product planners and engineers that the use of aerodynamics required certain principles to be respected throughout the car's design. If this breakthrough had not been made, it must be doubted whether Ford would have been able to move on to the near-revolutionary approach of deciding a car's shape early in the development process, as an integral part of its functional engineering.

At the same time as the new Escort was under development, Ford of Europe was working on a replacement for its ageing line of light- and medium-weight trucks, the 'D Series'. With Mercedes and other competitors moving away from the traditional generation of harsh, aggressive-looking vehicles, it was obvious that Ford should proceed in the same direction. But its solution, christened 'Cargo', was all its own: a slim, elegant-looking cab with unparalleled all-round vision which was so well designed that the company was able to abandon its original plan of building a second type of cab for heavier trucks. Ford was also able to use it on the Brazilian-made 'world truck' which it launched in 1985, with the US and Brazil as prime target markets.

The two key design factors behind Cargo's appeal were driver convenience (which was sufficient for even long-distance truckers) and the provision of enough airflow under the cab floor to cool even quite a large engine; this is an extremely tricky design task if the cab floor is to be kept virtually flat, as town delivery operators demand. In the years following Ford's decision to merge its European truck business with those of Iveco in the late 1980s, the Cargo remained a stalwart part of the combined product line.

From bulky aggression to slim elegance. Top: 'D Series' truck, 1965. Bottom: 'Cargo' truck, 1981.

The Cargo had cost £125 million to design, develop and tool. Ford of Europe's next project, the Sierra medium-sized saloon car ('sedan' in American terminology) required more than five times that financial commitment, including £95 million for design, development and engineering alone.

Cost was by no means the only measure of the importance to Ford of the Sierra, the replacement for the company's 20-year-old line of medium-sized cars, known in Britain as the Cortina and in Germany as the Taunus. It was the first vehicle to be developed from start to finish under the company's new marketing and design strategy.

The Sierra was adventurous indeed. In place of the blandness of the Cortina/Taunus came a vehicle which simply reeked of technology, drama and excitement. With an even more radically aerodynamic shape than Citroën's famous DS series (fondly nicknamed 'the platypus'), and its subsequent CX, it was not surprising that Lutz and Bahnsen again took time to win the support of the product planning and development committee, which initially favoured some of the less radical preprogramme alternatives. Together with the research engineers, Bahnsen also had to lobby for the adoption of some of the more ambitious aspects of the design, notably a large, injection-moulded polycarbonate bumper which was crucial to the car's low nose and ambitiously aerodynamic airflow management: the Sierra's 'drag factor' of 0.34 was a remarkable 24 per cent better than its predecessor's.

Fully realizing the risks of the dramatic step it was taking, Ford resorted to ingenious methods to soften up the market. During the 12 months before the Sierra's launch in late 1982 it put a very slightly altered version on display at several European motor shows, under the name of 'Probe III'. Hints were then dropped to the motoring press that this futuristic vehicle, totally unlike anything Ford had ever produced, might – just might – be the replacement for the Cortina/Taunus.

All the same, the Sierra had a mixed reception in Britain, where much of the car market is dominated by conservative company fleet buyers. Though the car's 'packaging' gave it a very spacious interior, its narrow front and curved lines made it look deceptively small. Some critics likened it to a cake that had gone flat at the edges, and others said it looked as if someone needed to take a bicycle pump to it. In the face of unprecedented competition, not only from the Germans and Japanese but also from a revitalized Vauxhall, the local General Motors subsidiary, its initial success fell short of Ford's expectations.

But in the crucial German market it sharply increased Ford's share of the segment for medium-sized cars. And in Western Europe as a whole it boosted the company's share of that category by half. At a time when the size of the total European car market was running 15 per cent below the forecast level – on which Ford and every other manufacturer had based factory capacity levels – this was not bad going. Competition in the European car market had never been so intense, nor profits so poor.

All the same, the Sierra's baptism demonstrated the risks Ford had run in being the first to lead the mass car market towards aerodynamic design. It is never easy or comfortable to champion a revolution. Another lesson was how difficult it can be to design one car to suit a set of regional markets, let alone the much-hyped 'global' market; it was the very aspects of the car which appealed to German buyers that deterred the more conservative fleet operators in Britain.

Within two years of the Sierra's launch, General Motors had begun to follow suit, first with a restyled version of its directly competitive model, the Vauxhall Cavalier/Opel Ascona, and then with a small Sierra lookalike, the Vauxhall Astra/Opel Kadett. By the spring of 1985, when Ford launched a larger aerodynamic car to fill the top of its range, called Granada in Britain and Scorpio in other markets, it was clear that it had succeeded in setting the trend for future European car design. With Peugeot, Renault, Volkswagen and others following the trend, the Sierra finally proved a major market success in the late 1980s.

Dearborn Follows Suit

Over in Dearborn, the parent company also swung into line. In the late 1970s, when the European Escort programme was already under way, Dearborn had joined it in a belated attempt to take advantage of the growing US demand for compacts and sub-compacts. But the top American management of the time had insufficient confidence in their European colleagues. They insisted on re-engineering the car from front to back, and top to toe. As a

Ford of Europe's most ambitious innovation: from the bland workmanlike
Cortina/Taunus (late 1970s, top) to the adventurous, high-tech Sierra, 1982.

result it grew more costly to produce, heavier, less fuel efficient and less satisfactory to handle. Sold under the names of Ford Escort and Mercury Lynx, it was reasonably successful, but made little money for the company.

Behind the scenes, however, the Dearborn designers had begun to press for the adoption of aerodynamics. They wanted to adopt a proper industrial design approach, in which shape was an integral part of the car's function, rather than a last-minute 'wrapping-up' of the working parts in a more or less stylish envelope.

So they were ready and waiting when Donald Petersen was named president in early 1980, at a time when Ford's market share and profits were plunging through the floor. In full knowledge of Ford of Europe's decision to launch the revolutionary Sierra, one of his first actions was to stimulate a last-minute rethink about the new US models under development at the time, which were scheduled for launch in late 1982 and early 1983.

As part of their newly revised strategy of abandoning full-frontal competition with GM and the Japanese, Petersen and his senior colleagues decided to target their attentions on particular market segments, especially the mushrooming hordes of young professionals ('Yuppies'). Starting with the Ford Thunderbird and Mercury Cougar in late 1982, and even more dramatically with the Ford Tempo and Mercury Topaz compacts the following spring, Dearborn broke away from the 'boxy styling offered by our domestic competitors', as one of its senior executives put it – a description which applied perfectly to Ford's own previous line of models.

Though theoretically committed to the idea of global products, Ford was unable to use the Sierra itself as a centrepiece of its American strategy. This was partly because it had decided, for several reasons, to design it with rear-wheel drive. Thanks in particular to the promotional efforts of GM, American consumers had been successfully convinced that a car of the Sierra's size needs to have front wheel drive if it is to be frugal on fuel. So Dearborn had to stretch the floorpan of the front-wheel-drive Escort as the basis for the Tempo and Topaz models.

Appropriately enough for the two cars which marked Ford's phoenix-like revival from the depths of consumer disfavour (and from a mammoth loss of over $3 billion between 1980 and 1982), Tempo and Topaz were first launched into the public's gaze on

Dearborn's first steps towards aerodynamic design. Top: Ford Thunderbird, 1982. Bottom: Ford Tempo, 1983.

the flight deck of an aircraft carrier which had survived the Pacific War with Japan – the USS Intrepid. The success of the two cars contributed heavily to Ford's remarkable feat in 1984, of lifting its US market share by a full 2 percentage points, to 19 per cent. By contrast, GM's share fell slightly, and Chrysler's rose by 0.3 per cent. (As already indicated in this chapter, Ford's share continued to rise throughout the 1980s, with GM's sliding the other way.)

By 1984 Ford had irrevocably committed itself to a strategy of integral design on its next two models. The ultra-aerodynamic Ford Taurus and Mercury Sable, 'mid-sized' twins which were launched in January 1985 after a record $3 billion development and tooling programme, were no reskinned affairs. This showed in their remarkable aerodynamic performance: a drag factor of only 0.29, the lowest of any US production car at the time, compared with 0.37 for the boxy cars they replaced, the Ford LTD and the Mercury Marquis. Fuel economy, passenger room and road performance were improved dramatically – car magazine writers were quick to compare them with Audi's high-status (and high-cost) imports from Germany. But the general public was most struck by their appearance. As *Fortune* magazine commented: 'these newcomers carry the jellybean look all the way – the front grille has disappeared altogether, and the hood curves down to the front lights'. By December 1985, with the new strategy already paying off in terms of market success, *Fortune* was commenting that 'the mainspring of Ford's US revival is its new boldness in design. Once a plodding follower of GM, Ford has moved aggressively to put its cars in the forefront.'

How Design Grew New Muscle

Behind this series of dramatic model changes in Europe and the US lay substantial alterations in organizational structure and product development procedures, all of them aimed at improving coordination between product planning, marketing, engineering and design.

As far as design was concerned, the most significant change in the US was the decision in 1980, as Petersen assumed the presidency, to move a large part of the corporate design staff into a line management role within Ford North America's product development structure. Under Jack Telnack, an American who had previously headed the European design team, the influence of the Dearborn designers grew by leaps and bounds. Along with Petersen, Telnack must take considerable credit for the success of the Taurus/Sable programme.

In Europe, a similar step had been taken as far back as the late 1960s. Design was put on a par with product planning and engineering, with all three reporting to a vice-president of product development. This elevated position in the organization structure, rare in any company until the 1980s, gave the European vice-president of design more influence than his American counterpart of the time.

Strategy and structure apart, one of the main sources of the designers' growing muscle during the 1970s was the decision to recruit a number of design engineers into the design centre. For the first time, this enabled the industrial designers to talk to the engineers in their own language: in the past the 'styling' department (as it had been called) had often been 'raped' by the engineers, according to the evidence of some of the engineers involved. But nowadays the design engineers carry out a full costing of the designers' work, and make engineering feasibility studies, to make sure it is viable in economic and technical terms. In certain other motor companies, the work of industrial designers is still confined to styling and the creation of concept models, which are completely changed once they are handed over to the engineers. Such is not Ford's way. By 1984 Uwe Bahnsen's 370-strong department contained nearly as many design engineers as industrial designers.

Until the beginning of 1985 Ford's European product development structure was highly departmentalized. No one below the level of the vice-president of product development possessed clearly defined authority to coordinate the work of the different departments. In the early stages of development the product planners tended to 'hold the ring' between designers, development engineers and production staff (product planning itself represented sales and marketing). But their grip tended to slacken as projects progressed, with the result that Ford experienced many of the classic problems of multiple hand-overs from department to department: details were changed, work had to be done again, costs escalated and valuable time was lost.

To overcome these difficulties, and create more cohesion, a matrix structure was introduced. The product planning function was absorbed into four new 'programme offices', one each for small, medium and large cars, and another for power trains (engines and transmission). Reporting direct to the vice-president of product development, like the heads of design and engineering, the four 'programme directors' were given clear responsibility for the coordination of development. In addition to their own direct staff of product planners and engineers, they assumed dotted line control for the designers and engineers, though both groups still report direct to their respective vice-presidents.

Rather than indicating any lessening in the influence of design, the change was welcomed by industrial design as a necessary streamlining of the development process, with design now very definitely accepted as an equal partner. There was more resistance from some of the engineering fiefdoms, however.

Ford went on in the late 1980s to establish an ambitious global 'network' of design and development, with regional centres in the US, Europe and Japan (through its Mazda associate), splitting responsibility for products, regardless of the nationality of their ultimate markets. Thus large-car platforms were developed in the US, medium-sized in Europe, and small in Japan. It will take until the mid-1990s for the success of this global development strategy to be proved.

A Market Research Revolution

Just like any other executive, the actual muscle of a design chief depends heavily on the informal power structure, and on his or her own powers of persuasion. A measure of the muscle which Uwe Bahnsen managed to amass soon after his appointment in 1976 is provided by the dramatic changes in market research which he advocated, and which were first applied during the later stages of the Escort's development.

In common with hallowed American practice, Ford of Europe had always placed heavy reliance on 'product clinics' – sessions at which competitors' current models were disguised and displayed alongside existing and potential new Ford vehicles, in order to be assessed by a sample of potential customers. The standard motor industry approach had always been to ask the participants to rate all sorts of details of the various models against each other, right down to which had the best bumpers, rear lights, arm rests and so on, as well as the most effective overall design.

Not surprisingly, the participants – whether dealers, fleet operators or drivers – tended to judge everything against the features of current vehicles. The result, as in so many narrow market research exercises in other industries, was support for only slight design improvements to the previous model, and resistance to any dramatic change.

These tests used to be accorded an almost religious significance, and taken as a precise guide to the way a vehicle should be designed. As Bahnsen complained, 'we asked them for detailed reactions in a most ridiculous way, and then used the results as a substitute for decision-making'.[76]

Starting with the last of the three clinics which was held during the Escort programme, Ford went over to a more conceptual approach, in which the questions asked in the clinics were focused more on people's general perceptions of what a vehicle should look like four or five years into the future. Various methods were used to try to get the participants to forget about what they have just been driving. One line of approach was to precede the clinic session with a so-called conditioning group, in which they were given an idea of how trends were moving in other areas of design, such as architecture, fashion and office equipment.

As subsequent experience has shown, this helps adjust the participants' eyes and minds to the fact that some of the vehicles they are about to see belong to the future, not the present. The subsequent list of questions concentrates more on their overall perceptions of the designs, though there are still some detailed inquiries about particular design features.

Most important of all, the results are no longer viewed as the holy grail by Ford's marketing staff and programme directors. The company has dispensed with its dangerous faith in the overwhelming power of market research, and has learned the true nature of marketing strategy.

9 A Global Quest for Speed: Philips

The top executives of Philips spend a lot of time these days puzzling over why consumers flock to buy shoes, shirts and jeans with 'designer' labels on them.

No, this rather Calvinistic Dutch mammoth, known for almost every conceivable kind of electrical and electronic product, from lighting to microchips, computers, shavers, hi-fis, TVs and telephone exchanges, is not planning to compete with Pierre Cardin or Calvin Klein. Nor does it intend to buy them, and add their names to the already voluminous list of brands by which it is known throughout the world: Magnavox, Norelco and Signetics in America, Bauknecht in Germany, Pye in Britain and Marantz in Japan, to name just a few.

But it has become obsessed with the need to match the skill of Sony, JVC and its other Japanese competitors in creating consumer demand for new products. And it sees design playing a crucial role in spotting likely consumer preferences: not only for colours, shapes and styles, but also for entire product concepts.

Until the early 1980s Philips had always been dominated by technology. Its engineers, considered the 'kings' of the company, worked hard to perfect each product that emerged from the laboratory, assuming that customers would automatically flock to buy them. As a result Philips products were often overengineered, overpriced, late to the market – or all three. Across much of its vast product range it began to suffer at the hands of speedier, more cost-efficient and more marketing-conscious competitors. In hi-fi and video recorders its failure to keep pace with the Japanese was especially acute, and its consumer electronics operations plunged heavily into deficit.

So no one was surprised when Philips' top management decided that much greater priority had to be given to marketing and industrial design. It began to streamline the organization's notoriously elaborate and slow process of decision making, putting marketing executives in sole charge of many key positions that had previously been occupied by a management duo – one technical, the other commercial – who had had to take every decision in tandem. And it began to make much greater use of industrial design, not as a sidekick of marketing but as a self-standing originator of ideas, as well as a synthesizer of other people's.

The design revolution at Philips began slightly before the elevation of marketing, rather than following in its wake, as in many other companies. The key step was the recruitment in late 1980 of Robert Blaich, an American who for many years had been design director of one of the most design-conscious US companies, the Herman Miller furniture concern.

Though Blaich's predecessor as managing director of the Concern Industrial Design Center (CIDC – the word Concern is synonymous with Group) reported direct to the Philips' management board – giving him high formal status – many of his staff had lacked authority in their daily work with the company's 12 product divisions. This was partly because their chief was considered by most executives to be a stylist, rather than a manager–coordinator, and partly because of the way the 200-person CIDC was organized.

One of Blaich's first moves to rectify this situation was to give the Center a more devolved structure, in which a designer's seniority was determined not only by his technical expertise and experience, but also by his forte as a manager. Six 'design managers' were created, each assigned to be responsible for the Center's work for particular product divisions, and with several contact designers reporting to each of them. They were urged to integrate themselves as closely as possible with the divisions.

The new structure encouraged divisional engineers and marketing executives to look on designers as like-minded and useful generalists, rather than as specialized and arcane technicians. In several divisions, it eased the designers' way onto product programmes at a very early stage, and even in two cases onto various general committees, including a marketing group and a task force examining 10-year forward strategy.

In parallel Blaich fought for considerable improvements in the design briefing process. When he arrived the designers were frequently being brought into new product programmes at a very late stage, and being given as little as two or three weeks to do their work. All they could accomplish under such circumstances was skindeep styling. By 1983 things had improved considerably. Even where the designers were not being brought in at the very start of the development process (and this was now in the minority of cases), they were receiving more information at an early stage. And they were being given eight months, at the very least, to do their work. This allowed plenty of time for collaboration and discussion with the rest of the development team, a crucial part of the process from which the designers had often been excluded.

This is not to say that the designers' gradual integration into divisional decision making was always an easy process. The reaction of the divisions to the prospect of design playing an equal part was not always positive; with development engineers tending to try to override the newly upgraded marketing executives and product managers, designers sometimes were caught in the cross-fire. Much depended on personal chemistry between the people concerned.

If some managers continued to view Blaich's designers as stylists, others grasped the opportunity to use them as a link between various specialist functions. A prime case in point was the hard-pressed hi-fi product group, whose chief quickly began to use Blaich's designers as a bridge between marketing and development engineering. He treated them as a direct resource, rather than just as members of the product team beneath him. Blaich himself became a leading member of a special multi-disciplinary hi-fi 'innovation team' that was established in 1985.

As part of the 'globalization' of the product development process and its transformation from a series of sequential phases into a set of overlapping ones, this close collaboration helped achieve a dramatic shortening in the development cycle for hi-fi products. A 'midi' (medium-sized) system launched in the summer of 1985 took only 21 months to develop, a good seven months less than its predecessor. On less complex products the cycle was cut even more sharply, from 18 months to only 10.

With every passing year, development times are being compressed still further. This is in spite of the increased complexity

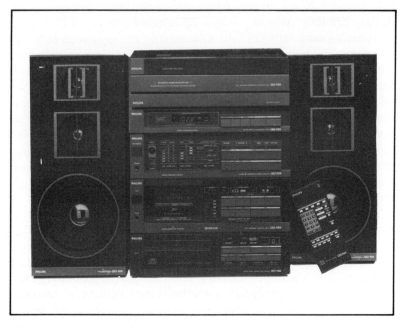

Developed in double-quick time in Philips' desperate duel with the Japanese: its 'Midi' hi-fi audio system, 1985.

which globalization has brought. Since late 1984 all new hi-fi products for Magnavox, Marantz and Philips itself, as well as its various other European brands, have been developed in common programmes. Despite the very different market segments at which the various brands are aimed – Marantz is a particularly prestige label – they are increasingly sharing key components such as circuit boards and tape mechanisms. This is an obvious way to improve economies of scale, in both development and production, but it had never before been done with Philips hi-fi.

Another initiative which has given design much more influence in the hi-fi product group, as well as in other parts of Philips' consumer electronics business, has been the creation of a series of ad hoc 'design research teams', to propose new product concepts and push them through the commercialization process in double-quick time. This mechanism has given birth to several up-market products which took Philips into new market segments.

The restructuring of the CIDC in 1981 enabled the industrial designers to start playing a catalytic role not only within particular divisions, but also between those whose technologies, products or markets were beginning to overlap or converge. One design manager was made responsible for all five divisions producing 'professional equipment' (medical systems, data systems, telecommunications equipment, scientific equipment and electroacoustics); another for both audio and video; and another for the two appliance divisions (large and small products).

One early outcome of this change was the discovery that three different divisions were planning to develop similar TV monitors; the designers persuaded them to cooperate and make only two. Another effect was the design of an advanced 'kitchen concept', embracing the products of the two appliance divisions. A third was the ability to develop programmes for harmonizing the graphics, appearance and overall corporate image of products in various divisions. The designers began with professional equipment, and then moved on, first to home entertainment (audio plus video) and then to large and small appliances.

It took Philips another four years to follow the example set by the CIDC, and integrate divisions whose activities had been converging for some time. From the beginning of 1985 it merged the separate audio and video divisions into one entity, 'consumer electronics'; a similar marriage took place between scientific and industrial equipment and electroacoustics, and telecommunications and data systems were also brought together.[77]

By this time Blaich had carried the reorganization of the CIDC a stage further. He had brought in considerable new blood and, in order to stimulate fresh ideas, had instigated a policy of frequent job rotation.

He had taken a tighter grip on the designers employed by the company's various subsidiaries in America, Japan, Australia and other faraway places. In divisions where designers had been located in several different places, often as isolated individuals, they were brought together to form a stronger joint force; in hi-fi, for example, they were centralized at one site in Belgium.

Such moves reinforced the ability of the designers to push their own product concepts into the pipeline, and to synthesize ideas which were welling up elsewhere. In the words of a design manager who worked for several years with the lighting division, before shifting to personal care appliances, 'marketing and engineering people mention things without realizing their potential. The design environment is a cacophany of noise'.[78]

One of his designer's innovations, picked out of the cacophany, was to suggest the combination of two sorts of recessed light fittings, in order to realize production output and gain economies of scale. After consulting the lighting division's marketing specialists, the design manager took the project to the development engineers. It was commercialized rapidly, and sold successfully.

A second product, which sold like hot cakes in an arresting piece of up-market packaging as 'The Light Point', reflects a different aspect of the CIDC's work. Through unofficial channels, a development engineer made contact with the same design manager about a revolutionary method of combining a halogen lamp with the necessary transformer, in place of the conventional approach of using a separate fixture to house the transformer. In the mid-1970s the status of design had been such that the engineer would never have been encouraged to come to the CIDC; he would just have been told to carry on alone, and design's only contribution might have been to put a graphic emblem on the side.

After developing the basic concept further with the engineer and his colleagues, the design manager then plunged into deep discussions with the division's commercial departments. Almost inevitably, they wanted something cheaper – as if price was everything in consumer purchasing decisions. But the designers' concept won through, and sold in hundreds of thousands at over $200 a time.

Among the lighting designers' other successes as product champions, catalysts or just go-betweens was the development of a series of modular fittings for spot lamps. This led to a commercial decision to segment the market differently, and gave Philips the immeasurable competitive advantage of being able to launch a wider range of 'spots' than ever before.

It was a similar story in the Small Domestic Appliances division. One of the most striking cases there was the 'Rotashaver' which the company launched in 1983, to considerable consumer acclaim. Even before the CIDC had been reorganized and the post of design manager created, the designer on this project had become so highly thought of by the divisional managing director, and his coordinating skill so openly recognized, that he had been made

Design-led innovations. Top: 'Light Point', 1980. Bottom: 'Professional Spot, 1985.

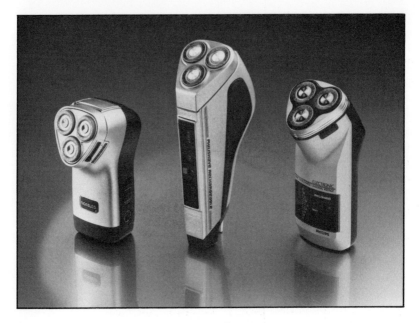

Three generations of razors: 1978, 1980 and 1983.

a member of its project team, and of several other management committees.

Development of the 1983 Rotashaver had begun as far back as 1978, two years before the company launched the previous model – its first shaver with twin 'retractor' blades on each of the three shaving heads, in place of the traditional single blades.

To communicate to consumers the technological innovation contained in the 1980 shaver, the marketing staff had wanted it to look different from its predecessors. So the designers provided a streamlined shape, with the blades not at 90 degrees to the stem of the shaver as in previous models, but emerging directly out of it. To power the extra blades, the 25-year-old motor design had to be 'stretched' and enlarged. The result was a heavy shaver which was awkward to use. But a portion of the market found it extremely attractive and it sold well.

The trouble was that, as consumer tests showed only too clearly, the rest of the market disliked it intensely. This made the divisional team nervous, since Philips has always aimed to give its shavers a very general appeal. This had to be restored if it was to avoid dangerous competitive attacks – including from the Japanese, who were already going for much smaller shavers, on the usual Japanese principle of precision miniaturization.

All this strengthened the designer's hand. Unhappy with the ergonomics of the 1980 shaver, he also wanted to get back to the size of the previous line of machines, or even below it. Given the need for extra power and extra transmissions for each of the three twin blades, this was a pretty tall order. Leaving extra room for all sorts of future electronic features, including automatic voltage selection, meant that the engineers would have to develop a new type of motor which was less than half the size of the one in the 1980 model.

The development engineers were understandably not amused, but with his marketing colleagues behind him the designer used sketches and models to win a series of arguments – and to justify the expenditure of nearly $15 million on the new motor line at a time when Philips' group profits were in a nosedive.

The designer also won a battle for the use of better quality plastic for the shaver's casing. Thermosetting, used for the previous 25 years, provided problems in production, breaking easily and coming out of its mould with rough edges. Thermoplast, which the designer advocated, gave much more precise surfaces and was easier to make in varied thicknesses.

Again the designer had to combat conservatism, in this case from production engineers in Philips' North Holland shaver plant. His weapons in a more general struggle to improve the quality of manufacture included chopsticks and Japanese bamboo, which he took into the factory to illustrate Japanese dedication to the meticulous treatment of materials. The effect? It was still a fight – almost as tough as over the motor.

The 1983 shaver was certainly a great improvement on its predecessor. A third smaller, and half the weight, it was altogether more satisfactory to hold and use. Under the brand name of Norelco Rotatract in the US, it gave a significant boost to Philips' market share. In Europe, under various names (such as Philips Rota in West Germany, and Philishave Electronic in Britain), it more than maintained the company's stronger market position.

The basic design laid the foundation for several other shaver products later in the 1980s.

Conclusion

Many of these examples are of lightweight consumer products whose technology is relatively mature and where, in the words of a member of the Philips management board, 'design is as much as 80 per cent of the product', in terms both of its market appeal and the key decisions in the development process.[79] But as the 1980s progressed industrial design began to find its way into nooks and crannies throughout the labyrinthine Philips organization.[80] It had very definitely become a vital strategic weapon in the company's fight for survival and supremacy.

10 Oddball Convert:
Baker Perkins

Michael Smith is a marketing man who for many years ran one heavy engineering company, Baker Perkins, and went on to head part of another: APV, the process machinery maker which merged with Baker Perkins in 1987. Smith is not the most likely enthusiast for industrial design. But ardent enthusiast he very definitely is.

Why does a manufacturer of machinery for baking, biscuit making, and – for many years – printing, use industrial designers *at all*, let alone give them a central role in the product development process? The products made on APV/Baker Perkins machines over the years have certainly been glamorous enough, from French croissants to glossy magazines, but the machines themselves have not been. So why did Smith and his cohorts of engineers and marketers become convinced that industrial designers were capable of strengthening the company's position so forcefully in world markets?

Baker Perkins' imaginative use of industrial designers first came to light in 1981, when Smith made a speech on quality and competitiveness at a major conference of top executives in London. To illustrate his case he took the example of his company's biscuit-cutting machines. The traditional ones had been very much geared to the home market: heavy and of solid construction. They were only occasionally sold overseas. But now, said Smith, Baker Perkins had a range of products which, despite being more expensive than their competitors, had gained a competitive edge in international markets because of superior performance. Its market share was growing and its margins were improving.

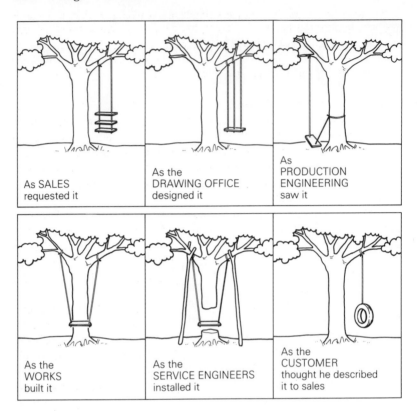

Mike Smith's secret weapon: the salutary tale of 'How not to design a swing, or the perils of poor coordination'.

To enforce the company's golden rule that the product, not the organization, is paramount, Smith and his technical director, Charles McCaskie, had created product management teams to bring marketing, engineering, finance and manufacturing together to define and execute a product plan.

Then came the comments that really surprised Smith's audience:

We gave the industrial designer a much greater role as the product planner, at a very early stage in the cycle. He no longer designed pretty guards to wrap our products in, but became the translator, the bridge, the catalyst. He turned the marketing specification of the product into reality before the design, the materials or the manufacturing methods had been established.

We now had a product range suitable for a world marketplace, with product quality the known responsibility of all the people involved. We had brought the marketplace into the factory. Product quality had beforehand probably had most to do with the field engineers, who kept dark secrets because no-one else wished to know.[81]

Without this organization, said Smith, Baker Perkins would have suffered from the notorious 'swing syndrome'. His set of slides, which gradually unveiled the perils of poor interdepartmental coordination, brought him considerable applause. Smith went on to recount how the company had produced a better product than its German and other competitors, with sales and market share to match.

Eighteen months after this first revelation of Baker Perkins' surprising commitment to industrial design, an even more graphic illustration emerged into public view: the story of its latest printing press, which at the time of Smith's speech had still been under development at the company's headquarters in Peterborough, 75 miles north of London.

The new press had emerged from a dilemma that had faced Baker Perkins as far back as 1976. At that time, its printing machinery division was selling just one type of web offset machine, known as the 'satellite' design. Its main competitor, Harris Corporation of the US, had backed an alternative type of design, the 'blanket-to-blanket' press. (Don't worry about the jargon, just enjoy the rich images it evokes.)

As 'blanket-to-blanket' technology improved, other competitors followed suit. So, especially in the US, where Baker Perkins was already well established, did the most rapidly growing and lucrative part of the market – not newspapers, with which the company had done considerable business, but the printers of magazines and brochures. As a result, the company had to scrap its range of half a dozen 'satellite' products, plus a major new development project. Instead, it embarked on a costly £2 million development programme designed to leapfrog the competition into the very top end of the 'blanket-to-blanket' market, with a machine that could print 32 pages at once, at a very high speed.

It was a bold, all-or-nothing step, and cost several times what Baker Perkins was used to; this is true even if one takes account of a 25 per cent grant from the British Government, which reduced the risk to measurable (though still uncomfortable) proportions.

Less than three years later, first orders for the new 'G16' machine were being delivered. This was in spite of all the intricacies of using computer-aided design and manufacture for the first time, and having to develop microelectronics expertise at a rapid rate of knots. The G16 went on to outsell the US competition, and to take about a fifth of the world market for these big machines, including 70 per cent of the North American market.

By late 1980, Smith, McCaskie and the management of their printing division had begun to get to grips with how to penetrate down-market with a slightly slower, 16-page machine. Not only were potential customers starting to invite them to do so, but they had to broaden the product range in order to reduce the division's vulnerability to a single product/market segment, and at the same time gain economies of scale from a range of modular products sharing as many common components as possible.

This is where the work of the industrial designers came in. It had been in the late 1960s, when he was working in the company's baking division, that McCaskie had hired his first industrial designer (from London's famous Royal College of Art). The job of the new arrival was essentially to do what McCaskie called 'clean-ups' on the machinery's casing. As McCaskie climbed the ladder – he was appointed technical director of the whole company in 1971 – industrial design grew with him. First the industrial designer was allowed to use his sketching ability to work on alternative product concepts, and then to go a stage further by making alternative models.

As McCaskie recalls, 'during this period we were developing marketing – and we realized that industrial designers also had a role to play'.[82] As more designers were recruited they became increasingly involved in specification setting. By 1980, when the specifications for the new press were first debated, they had become fully involved in the development process. This is now the standard pattern: together with a senior design engineer (the team leader) and a production engineer, one of the company's half-dozen designers forms the core of each product team. Marketing, sales and service provide input, and are involved in decision making, but are not part of the three-person core.

The industrial designer on both the printing press projects was Martyn Wray, a former church organ builder and subsequently a student at London's Central School for Art and Design. In a book about the work done for manufacturing industry by former Central graduates, Wray summarized the pivotal role of the industrial designer in modest fashion. He had particular responsibility for appearance, ergonomics, hygiene and safety, he wrote, and 'his sketching ability is also used as a link between members of the design team, other parts of the business and the customer'.[83]

McCaskie sees the industrial designer's role very much in terms of the classic comment that engineers are good at analysing, industrial designers at synthesizing. But he also goes much further, stressing the extreme importance of the designer as communicator: the communications skills of industrial designers are of an entirely different order of magnitude from those of the engineers, he says. Himself a mechanical engineer, he argues that engineers tend to focus on cogs, wheels and mechanical forces, rather than on the product as a whole.

It is not just engineers and marketing people who don't speak the same language, but also the design engineer and the production engineer. In McCaskie's words, 'a manufacturing guy doesn't understand the process of printing at one speed rather than another. It's not part of his culture'. Industrial designers are needed to help the two communicate.

On the second printing press project, christened the G14, Martyn Wray began synthesizing the analysis of his various colleagues (and adding some of his own) in early 1981, right from the start of the all-important specification phase. It was through specification writing that the industrial designers had really begun to be influential at Baker Perkins. To synthesize the specifications of marketing, engineering and production can be an appallingly difficult task, which is why many companies find that their new products are difficult and expensive to make. Or they prove easy to manufacture, but fail to meet consumer requirements. It is equally possible, as the 'swing syndrome' shows, that the various sides settle on a weak compromise which suits nobody.

During the specification phase of the G14 Wray played a key role in challenging the proposals of several specialists, including some which concerned the question of which market segments to go for, and therefore precisely which performance characteristics to specify. And during the subsequent 12 weeks, when the product team was trying to establish whether it would be possible to make the new press to the extremely tough cost targets which had been agreed (30 per cent cheaper than the G16, on a comparable basis), Wray's sketching ability proved vital in establishing a basis for the intricately detailed communication which had to take place between the various parties.

His imaginative proposal of several new concepts for certain components was also of considerable assistance in getting the cost down. Most engineers find it difficult to sketch – they're not trained to – and their lack of practice at applying their imagination often makes them miss even simple solutions.

And so the catalogue of industrial design virtues at Baker Perkins continues, right down to the 'bottom line': their ability to help shorten the product development cycle not just indirectly, by helping specialists communicate with each other, but also directly, by accomplishing through a few quickly done sketches what would take an engineering draughtsman weeks to do, even with the help of CADCAM. The entire development-to-delivery process of the G14 took only two years, a third less than it would have done in the late 1970s.

Added to this contribution was some more obvious but nevertheless important work on aesthetics and ergonomics, including requirements for easy field installation and maintenance, plus the resolution of some very tricky safety aspects to do with access to the top of the press.

The first G14 was installed in Chicago in early 1983, and within two years over 30 had been sold, including 200 printing units – which is a large number in the printing market. Along with continuing sales of G16, the success of the machine helped to quadruple turnover of the printing division to £40 million in 1984, and to boost profitability dramatically.

Following the Baker Perkins merger with APV, however, the printing division became something of an oddity, outside the enlarged group's main focus of process machinery for the food and beverage industries. As a result, the division was sold in 1989 to one of its main competitors, Rockwell International.

But some of the Baker Perkins industrial designers – still located at their original Peterborough base, though now within various of the APV group's operating companies – continued to work on a consultancy basis for the printing division under Rockwell. The successful development of a new printing press, which was launched in 1990 as the G-44, again owed much to the designers' involvement.

So, within APV, did two products also launched in 1990, a flaking roll for breakfast cereal production, and an innovative skinless sausage-making machine. In both cases, Martyn Wray was again one of the industrial designers involved.

If a former church organ builder can play a crucial part in dismantling departmental barriers within a capital goods company, and can help it improve its commercial performance, then anything is possible. The common denominator is industrial design.

11 The Designer at Work:
Kenneth Grange of Pentagram

From his studio in a converted London dairy, Kenneth Grange cannot stray far without confronting his own creations. A short step to the east lie three great railway termini, bustling with the High Speed Trains whose power cars he designed. Further towards the centre of the city, along streets lined with his parking meters, are ranks of department stores displaying his food mixers, irons and shaving equipment. Outside his native Britain, too, his products are pervasive. The Scandinavians go for his Kenwood kitchen appliances and the French, Germans and others have flocked to buy his innovative Wilkinson Sword razors. By 1990 much of his work was in Japan, where his sanitary ware (bathroom suites etc.), gas appliances, cosmetics packs and spectacles were being bought in their thousands by local consumers.

Grange, a tall, engaging man with well over 30 years' experience, is the industrial design partner at Pentagram, one of the world's leading design consultancies and one of the first from Europe to set up in the US (in 1978). His co-partners include not only a bevy of top international graphic designers, but also a well-known architect. The client list of the firm reads like a who's who of international business: Hilton International, ICI, Kodak, Mitsubishi, Nissan, Penguin Books, Pirelli, Reuters, Shiseido, Time-Life, Unilever, Warner Communications and many more.

Pentagram's work includes not only the bread-and-butter of design consultancy – one-off contracts which are known in the business as 'jobbing' – but an unusual number of much more far-reaching assignments: several partners act as 'consultant design directors' for clients such as the prestigious publisher Faber & Faber. Grange himself has held several such positions, notably at Wilkinson Sword and part of the Thorn–EMI group.

In their impact on some of these clients, the Pentagram partners have been just as influential as the great American consultants of the past, such as Henry Dreyfuss at Deere and Eliot Noyes at IBM. Not only have they helped guide their clients towards greater commercial success, they have instilled design management principles and procedures which are now seen as models of their kind.

At Wilkinson, Grange had a particularly strong impact through-out the 1980s. He played a very full part in its product development process, creating function as well as form, and thereby giving Wilkinson a technical edge for part of the 1980s over its long-standing arch-rivals Bic and Gillette. Through his position on several senior marketing and product development committees, he also had an influence on the company's marketing strategy. And he acted as a cohesive force between different functional specialists and geographic units which previously held themselves proudly apart.

Contrary to the conventional image of the egotistical and individualistic designer, Grange shows a remarkable modesty about his many years of professional achievement. It is this very self-effacement, combined with a quiet sense of humour, which has frequently helped him overcome managerial resistance to design, and which has made him such a valuable team member within so many client companies.

Trained at one of the many London art schools just after the Second World War, and then as an architectural assistant, he moved on to work for several small design consultancies. He quickly set up on his own and in 1958 received his first big commission – the face-lifting of a parking meter. Closely on its heels came a premier international contract – from Kodak. His relationship with Kodak's European offshoots lasted until the late 1970s, when Eastman Kodak decided to centralize design at its in-house studios in the US. In the meantime Grange had helped put Kodak's European camera operations well and truly on the map, with the Instamatic range in 1958 and the Pocket Instamatic in 1975. Both products were remarkably successful, bringing low-cost photography to millions of people all over the world. The only

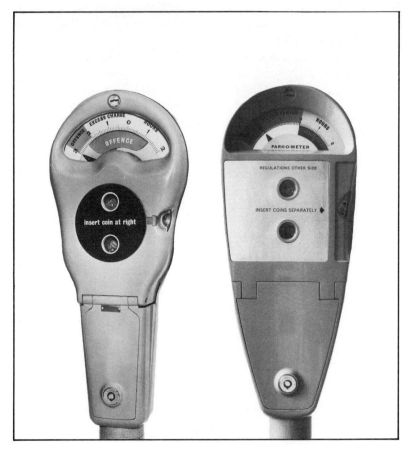

Grange's first big commission – a facelift of 'Venner' parking meters in 1950. Before (left) and after.

areas in which Grange's design was not sold were North and South America, which were supplied with US designs.

To the extent that he was not involved in the design of the film cassette, the lens or the exposure system, Grange's work for Kodak might seem 'mere styling'. But it was more than that. He not only designed the correlation between trigger, winder and grip, but also the shape of the product – which in a camera is a highly functional matter: it must be comfortable and stable to hold and use. As Grange himself says, in such cases 'it is unclear where style starts and ends'.

Instant successes with European consumers: Kodak Instamatic cameras, 1958 (top), and Pocket Instamatics, 1975.

His next big contract came in 1960, from the Kenwood appliances company, which at that stage was still managed by its founder, Kenneth Wood (it was later acquired by Thorn, and in the late 1980s became a management buy-out). His first product for Kenwood, the Chef food mixer, was so successful that a redesign was not needed for 15 years. His second, in 1966, was intended to be a straightforward redesign of a small hand mixer, the Chefette. The company told him that it might want to offer a stand and bowl as accessories at a later stage, but not yet. Grange decided to

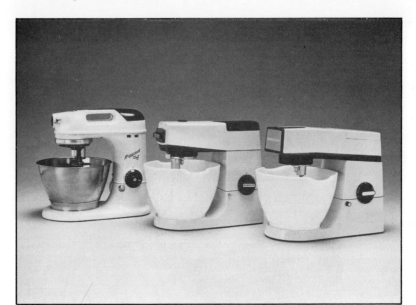

Three generations of Kenwood Chef food mixer. From left: pre-1960, 1961 and 1975.

integrate the stand and bowl from the start, however. When the work was done,

> without referring to what I had actually been asked to do, I showed the whole thing. It was very well received, and no mention was made of the stand and bowl that had not been asked for. The moral here is that regardless of how perceptively a brief is written, the designer is employed to be creative; and providing he meets the brief, he must always have the audacity to push an opportunity that bit further.[84]

Grange has been repeatedly ready to go further than his clients have envisaged, thereby extending his role from styling into functional design. For Maruzen, a Japanese sewing machine company, he proposed the design of a really small, lightweight machine with short drive shafts. The Japanese engineers were sceptical about various technical aspects, but eventually agreed – and the machine became extremely successful in both Japan and Europe.[85]

The Kenwood Chefette, 1966. Grange exceeded the brief – successfully.

Grange again exceeded his brief when, shortly after merging his practice with another to form Pentagram in 1972, he designed the power cars at each end of the High Speed Train. The diesel-engined 'HST', as it is familiarly known, went on to form the backbone of British Rail's sleekly modernized Inter City network of nationwide passenger services. Thanks to nightly trials in the wind tunnels at London's Imperial College of Science and Technology, what had been intended as merely a stylistic shape was given an unexpectedly impressive aerodynamic performance.

A key factor in the achievement of the best possible airflow and lowest possible fuel consumption, given the specification for rapid acceleration to 125 mph, was Grange's initiative in designing a cab without buffers; this was a very radical step for any group of railwaymen to approve, and they took some persuading. In many ways, Grange's contribution to the HST echoes the achievement of the European locomotive engineers of the 1930s, who paid far more attention to marrying beauty and performance than did their American counterparts, for whom style was the overriding criterion (see chapter 2).

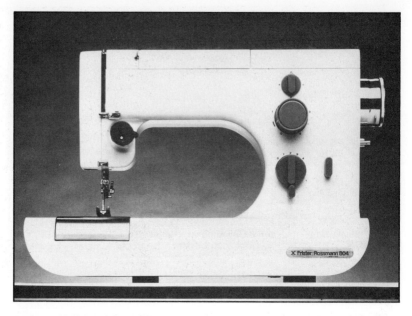

Japanese-made sewing machine, late 1960s. Grange had a key influence on its convenience for the user.

One of Grange's great strengths in his collaboration with his clients' engineers is the patent enthusiasm he demonstrates for the minutiae of manufacturing. This shows itself in all sorts of ways, whether he is recalling the extemporizing which often occurs (he used a vacuum cleaner in his father's garage to spray-paint an early model of the Kenwood Chef), or in his description of engineering toolmakers as 'the aristocrats of production engineering', and their work as 'beautiful and fastidious'.[86]

In spite of his expertise at combining form and function, Grange makes no apology whatever for the importance of appearance in an industrial designer's work. He describes as 'one of my hobby horses' his view that 'the industrial designer's role is to make products more of a pleasure to use'. This includes making them more attractive to see, to touch, to hold and (where applicable) to hear and to smell. All the better, he says, if, as part of this process, he can make them perform 'differently' (he hesitates to say 'better' for fear of denigrating the engineers he admires so much).

British Rail's 125 mph High Speed Train, early 1970s. Grange's initiative improved its aerodynamic performance.

As befits someone who has worked for several domestic appliance manufacturers who have been heavily influenced by the quality and character of Dieter Rams' designs for Braun products, Grange is a great admirer of the Braun tradition of closely combining form and function. But he distances himself from Rams' ascetic view of design: that a product's appearance should invariably be as simple as possible, with no unnecessary trimmings and hardly a bright colour to be seen.

Grange certainly favours what he calls 'declared simplicity' in domestic appliances, as opposed to some of the illogical and 'glittering confusion' which has characterized the design of many products, especially in America. But he stresses that 'I do believe in a bit of cheerfulness and glamour'. The traditional European design adage of 'fitness for purpose' can be taken to extremes, Grange feels. He disagrees with the argument that products should be stripped of everything superfluous to their function, and says that 'part of the purpose of a product is to give pleasure. That is one of the designer's key contributions.'

This certainly applies to Grange's work for Wilkinson Sword, which was one of the main offshoots of Allegheny International, the American steel-to-consumer durables conglomerate for a number of years before being sold to Swedish Match and then, in 1989, finally agreeing to submit to the tentacles of a hungry Gillette. After Grange resumed work in the late 1970s for Wilkinson, having carried out a few 'jobbing' contracts in the 1960s, the group's razors, scissors and garden tools became much more attractive to look at, as well as more satisfactory to use.

Wilkinson, an old English family firm whose original business was the making of swords, first hit the international scene in the 1960s when it produced the world's first coated razor blade. Especially in Europe, but also elsewhere, this made a dent in Gillette's dominant position in the market for wet-shaving equipment. A second success, in 1971, with 'bonded blades', was short lived: the twice-bitten Gillette fought back with a vengeance, giving an aggressive promotional push to its own technological innovations (double blades, and swivel-headed razors). Then, in the space of just two short years, both Gillette and Wilkinson were sent reeling by the smash-hit success of yet another new type of product: the disposable razor, made available to an eager world by Baron Marcel Bich, under the brand-name of Bic.

Gillette managed to fight back, albeit less successfully than against Wilkinson. But the British company slipped behind, preserving a respectable market position only in the UK and Germany. Though caught in a vicious price war, it restarted its efforts to create another technological breakthrough in the research laboratory.

It was only after the company was acquired by Allegheny in 1980, and a bustling marketeer was put in charge, that Wilkinson began to recover its confidence as a marketing-driven organization, and see itself again as a mass-market supplier. In addition to making a number of new marketing appointments, one of the first actions of the new chief executive was to set in train the development of an ultra-low-cost disposable razor, using conventional technology. Grange was to play a central role in its creation – and success.

As part of the previous strategy of serving premium market segments, Grange had already been rehired on a jobbing basis to work on the development of a very high technology razor. This

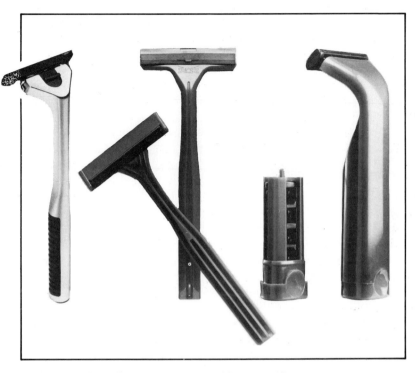

Razors which helped put Wilkinson Sword back onto the competitive map.
From left: Royale, 1977; Retractor, 1983; Hawk, 1985.

project was shelved, and Grange instead designed an up-market steel razor which used bonded blades. With the shift in strategy, he was also engaged to assist with the low-cost product.

From the very start of the mass-market project, which cost £3 million in development and tooling, he was given a full part in the development process. He had an influential voice in the company's decision to create an entirely new product, and use new tools and assembly machinery, rather than to adapt existing designs and machines. Together with the development engineers, he was also very much responsible for the product's key competitive advantage.

Many markets round the world insist on the blade in a razor being covered by a shield. The usual solution has been a small plastic cover. To save the cost of that extra part, the design team

combined the handle and the cover into one moulding, and created
a device which enabled the blade to be snapped forward for use,
and then retracted afterwards. This, and the ingenious construction of the sculpted handle, allowed Wilkinson to cut
manufacturing costs to the bone, and to price the product – called
Retractor – squarely in competition with Bic and Gillette, at just
a few pence (or cents) per razor.

With its bright red colour adding that touch of glamour which
Grange likes to give his products, Retractor certainly stood out
from its competitors on a supermarket shelf, or in a drugstore's
display bin. But the most important factor behind its success –
sales were well above forecast – was probably the way that its
innovative design reinforced its market positioning. It really did
achieve Theodore Levitt's test of 'meaningful distinction'.

By the time Retractor was launched in 1983, Wilkinson had
recognized Grange's strategic value in two significant ways: he
had begun to report directly to the chief executive, as well as
continuing to liaise with the head of development, and he had
been appointed consultant design director. With the chief
executive repeatedly emphasizing to his managers the role
industrial design can play in creating real differentiation, this
immeasurably reinforced the influence which Grange had already
gained by his own efforts lower down the organization.

His impact went on to be felt at various levels. Alongside the
chief executive and the heads of Wilkinson's key national
subsidiaries, he became a full member of the main marketing and
product development committees. Along with less senior marketing and engineering executives, he also sat on the new product
development groups which oversaw actual development work in
each of Wilkinson's business areas, notably shaving, garden tools,
hardware, knives and kitchen equipment.

As Wilkinson moved in the early 1980s towards the development of global products, Grange several times played a catalytic
role in creating cohesion between the often conflicting new
product concepts of subsidiaries in different countries. One such
case was in the design of the first unified line of garden pruners
for Wilkinson and Allegheny's American garden tools manufacturer. Grange's analysis of the possible range within which trade-
offs could be made between product features and price became

known as 'the Grange diagonal'. It was a marketing tool, supplied by an industrial designer.

Grange's impact on marketing strategy is further illustrated by a new type of razor which, without any initiative or 'brief' from marketing, was conceived in collaboration with the head of development engineering. Their idea for a 'semidisposable' product, with three blades in a cassette hidden in its handle, created an entirely new market segment. Initially sold in plastic as the 'Hawk', it was later developed into a metal-bodied product, the 'Compact', which sold well in Continental Europe.

In his design work for Wilkinson, Grange made some use of his own staff at Pentagram. But he also acted as a top-level advocate and channel for Wilkinson's own small in-house team of industrial designers. Except in the UK garden tools division – and then only sporadically – they had been underexploited before his arrival on the scene.

The Wilkinson of the 1980s therefore enjoyed a similar balance to Olivetti in the use of internal and external designers, but by different means. Olivetti has secured a 'window' on the outside world, and through it access to a seething mass of fresh ideas, by supporting two outside design studios. But it effectively reaps the benefit of an in-house resource by committing the studios to spending half their time on Olivetti projects. Most of the other companies featured in this book have a different approach. Deere relies solely on the Dreyfuss consultancy for its industrial design (it has virtually no in-house resources), while Sony, Philips, Ford and APV/Baker Perkins work almost exclusively in-house.

Grange is not dogmatic either way. He argues that the ideal arrangement may be entirely in-house, entirely external or a mixture of the two. It all depends on the traditions of the company; on its commitment to design; on the strengths or weaknesses of its in-house staff; and, above all, on the personalities involved. Just as industrial design will lack influence if the chief executive and the marketing function, in particular, are not committed to it, so a senior in-house designer will be ineffective if he or she is not a good and persuasive manager.

But heavy reliance on design consultants can only work if, like Olivetti, Deere and Wilkinson, the client appreciates the level of financial commitment required for the consultants to attend all the meetings that are necessarily involved in the planning and execution of product development.

Grange does not entirely share the complaint of many design consultants about the tendency of some clients to see a consultancy's work as 'jobbing'. If design is to play an influential part in the company's daily life, it must indeed be a continually available resource – whether in-house, external or a cross-breed of both. But the client may also have plenty of individual projects which call for short-term jobbing contracts; this applies particularly to product and graphic design, less to corporate identity work.

Despite his wealth of experience and his elevated position within so many client companies, Grange is still quite happy to spend part of his time as a 'jobbing designer'. He argues that the freshness that such work provides can help reinforce the designer's breadth of vision, as well as that fundamental designerly attribute: what Grange calls 'a sort of wilful naïvety which constantly challenges the way things have been done before'.

Part III
Globalization: Threat or Opportunity?

Introduction

Just like the word 'design', there is no single accepted meaning of the term 'globalization'. To some people, it means the globalization of *industries:* the growth of global competition between fiendishly clever companies which tend to vary their products from one country to another, to suit the taste of different national consumers. To others it implies an inexorable shift towards global *products.* To some, but not to others, it also embraces a shift towards standardized global *brands.* To some people the term describes a truly global homogenization, from Nashville to Nigeria, Stockholm to Singapore. To others it covers only Europe, the US and Japan. The permutations of meaning are confusing, not to say bewildering. To most people the only certainties are that globalization has become fashionable, and that it presents a daunting new challenge of indefinable proportions.

Chapter 12 attempts to dispel some of the fog by examining the reality beneath the hype. It suggests that the concept of globalization, though undeniably powerful, has certain limits. These have important implications for corporate strategy and marketing, and also for the role of industrial design: as chapter 13 argues, an overliteral application of the gospel of globalization could turn back the clock to the bad old days of skin-deep styling. Yet, combined with the brave new world of computer-aided design and manufacture, globalization actually offers an opportunity (and a justification) for the design dimension to be strengthened, not weakened.

12 Global Hype and Reality

There could hardly be a less glamorous herald of a new era in international business than a bright orange plastic container full of liquid detergent.

Yet this mundane product from Procter & Gamble broke new ground in several respects when it was launched in the winter of 1983–4. Not only did its development team include staff from the US and Japan, but it was launched virtually simultaneously in both countries: in the US as 'Liquid Tide', in Japan as 'Bonus 2000'.

In the past, in common with most other companies in its field, the products which P & G internationalized were handled in a much more cautious fashion. Items such as its 'Pampers' disposable nappies were designed initially with one market in mind, usually the US. If, like Pampers, they proved themselves successful over a lengthy period, often as much as two or three years, they were introduced abroad in a step-by-step manner, sometimes under different brand names and often with adaptations for local market conditions.

Even if its brand name varied between countries, the new liquid detergent represented one of P & G's first steps on the road towards globalization: the development, production and sale of products on a near global basis.

All over the US, and to a considerable extent in Europe, companies in all sorts of industries had begun trying to work the same trick by the mid-1980s: to gain the economies of scale and marketing initiative which the Japanese so dramatically grasped during the 1960s and 1970s in cameras, motorcycles, cars and

Losers and winners in the race to globalize: Nabisco's Oreo cookies flopped in many countries outside America, but Procter & Gamble's Pampers and Liquid Tide fared much better.

consumer electronics, and which had always been one of the secrets of success at a handful of Western companies – market leaders as varied as IBM, Coca-Cola, Levi Strauss and McDonalds.

What most Western companies failed to spot was that, by this time, their Japanese competitors had moved beyond the approach of 'globalize everything' to a much more sophisticated strategy of greater globalization of some products alongside increased national (or, at least, regional), differentiation of others – albeit with common components and other constituents wherever possible.

It was a pity, for millions of frustrated customers as well as for many corporate profit statements, that most Western companies took so long to realize that the essence of management in this age of globalization is not to have one simplistic strategy, but a thoroughly ambiguous mixture of several (this point is dealt with at greater length later in this chapter).

Not surprisingly, this new Western bid to create global products across a broad front had mixed results. It was highly successful in computers, electronic office equipment and other new categories of product where there were no existing cultural barriers to overcome. It also brought rich rewards in some apparently more 'culture-bound' markets, such as soft drinks (Seven-Up and Schweppes are notable examples).

But failures abounded in foodstuffs, Nabisco's chocolate sandwich Oreo cookies, which are phenomenally popular in the US, failed to appeal to European tastes and proved a dismal flop. In cars, neither Ford's nor General Motors' attempt to go global in the early 1980s was successful: Ford's Escort and GM's 'J-car' (known in Europe as the Vauxhall Cavalier or Opel Ascona) not only ran up against consumer resistance and internal corporate barriers, but also suffered from the difficulty of replicating locally the cost and quality levels of European component manufacture.

Yet the drive to create global products is certainly set to continue. Back in 1984 the hitherto secretive Procter & Gamble put shyness aside and allowed one of its top executives to confirm the trend.

We've progressed from national economies to an international business arena. Twenty years ago the introduction of a new product in Japan or Singapore wouldn't have created much of an impact, if any, in Germany, the UK or the US. But these days a product introduction in one part of the world is likely to be picked up, and emulated, anywhere else.[87]

With the launch rate of most types of new product accelerating in response to technological change, intensified competition and other factors (see chapter 4), P & G has not only had to design products to be global from the start, but to cut development times. Xerox, IBM and Philips have all discovered that the same pressures apply to almost the whole range of electronic products, while Ford and GM have tried to find ways to respond to Japanese competition by shortening their car development cycles.

'If we don't think through the concept of a new product on a global basis from the very start, we aren't able to enter additional markets until we see the results from the country where it is first introduced', commented the P & G executive. In his company's experience, global development, 'almost certainly saves years'.

One of the most succinct and elegant descriptions of the forces behind the growing globalization of industries and products – and, to a lesser extent, brands – was contained in *Triad Power* by Kenichi Ohmae, head of the Tokyo office of McKinsey & Co.[88]

(The triad of the title is Japan, the US and the European Community.)

Ohmae cited four main factors:

(1)   The growing capital-intensity of manufacture, which – until the widespread arrival of flexible manufacturing systems that allow cheap, short production runs – favours even larger economies of scale than in the past.

(2)   The accelerating tempo of technologies: the cost of R & D is soaring, new technology is diffusing through the industrialized countries more rapidly than in the past, and technological advantage is therefore becoming increasingly hard to gain and then sustain. As a result many companies are having to try to start penetrating the 'triad' with new products simultaneously, rather than steadily on the old country-by-country pattern.

It is these two sets of forces, in particular, which have prompted the now familiar and endless spate of new joint ventures, consortia and cross-supply links within the triad. They criss-cross such diverse industries and technologies as:

- aero engines (General Electric–Rolls Royce; Pratt & Whitney–Kawasaki–Rolls Royce);
- cars (in both components and assembly, linking GM with Toyota, Chrysler with Mitsubishi, Volkswagen with Nissan, VW with Ford, Volvo with Renault, and just about everybody in some way or another with everyone else);
- consumer electronics (Matsushita–Kodak, JVC–Telefunken–Thorn and even Philips–Sony);
- computers (AT & T–Olivetti, Hitachi–Hewlett-Packard, Fujitsu–Amdahl–Siemens–ICL, even IBM–Matsushita).

In many of these relationships, the Western partner is in an inferior position. A large proportion of the deals have been arranged to plug yawning gaps in the Western company's product range. But they almost invariably enable the Japanese company to use its partner's distribution networks as a low-cost way of testing the market, and of building a reputation with dealers and consumers. It then breaks loose to sell under its own name. There is nothing uniquely Japanese about these tactics. They are endemic to almost any licensing/OEM/joint venture relationship. European companies used to complain about Americans doing precisely the same thing. But the problem has become particularly acute with the emergence of Japanese prowess in so many sectors of industry.

One of the few means of defence for the Western partners in such deals is to use the relationship as a breathing space to develop, and come bouncing back with, the next generation of products. General Electric tried to do this in robots, as did General Motors in small cars.

(3)   Ohmae's third explanation for the globalization of industries and products was the emergence of a growing body of universal users on a much more mass market scale than in the past, when only the very rich could afford standardized global products such as diamonds, Sèvres china and Yves St Laurent dresses. The new 'global consumer' has emerged not only among the younger generation, in fashionable consumer goods such as Coca Cola, Pepsi, Levis, McDonalds – and, of course, much pop music. He or she is also increasingly appearing among professional users of engineering products, ranging from Caterpillar and Komatsu earth-moving equipment to Ericsson public telephone exchanges. The much-vaunted power of the new communications (mass travel, containerization and most recently satellite broadcasting), is yet another factor behind this trend.

(4)   The emergence of neoprotectionist pressures, which is forcing multinationals to attempt the extraordinarily tricky balancing act of becoming what Ohmae calls 'true insiders' within each country, while at the same time going global in product development and production.

Managing Ambiguity

But does the trend towards globalization necessarily mean, in the words of Theodore Levitt, that 'the world's needs and desires have been irrevocably homogenized', or, indeed, that they are becoming so? And does it mean, as he has argued, that the days of national and even multinational companies are therefore numbered, and that the future lies only with truly global companies which see the world as their home market, and which act accordingly?

Equally to the point, does it also mean that multinationals need to go global in every aspect of what McKinsey calls 'the business system' (research, design and development, materials sourcing, component supply and manufacture, assembly, marketing, sales and after-sales service)? Taking literally the message of Levitt's manifesto-like article 'The Globalization of Markets'[89], many companies initially thought of doing just that.

In most cases, they were wrong. It is quite possible to be in a truly global business where large-scale development and production is critical, but where marketing and advertising need to be tailored to individual regions or countries. Canon's AE–1 camera is just one example of a standardized product which has allowed plenty of room for differentiated marketing strategies.

McKinsey, which had been preaching the global gospel long before it became fashionable in the early 1980s, has repeatedly advised companies to distinguish most carefully between different product categories when they decide whether, how far and how, to go global. The McKinsey analysis reinforces, and is reinforced by, the extremely influential work of Michael Porter, a Harvard professor of general management whose books have had an even greater impact on American business since 1980 than Levitt's.[90]

Substituting the concept of the 'value chain' for McKinsey's 'business system', Porter argues that the competitive advantage to be gained from global coordination varies not only among industries, but within a given industry, and among stages along the chain. In insurance, for example, only a few specialized segments such as marine insurance are global. In TV sets, on the other hand, the bulk of the market, consisting of portable table models, is global, while more specialist segments are not.[91]

In a phrase coined by one of Porter's Harvard colleagues, John Quelch, global marketing does not necessarily mean providing the same product in all countries, but offering local adaptations around a standardized core. Global marketing may well require a standardized marketing *strategy*, but the programme for *executing* that strategy can vary from market to market.[92]

Nor does global marketing mean that all companies with narrow national markets are terminally vulnerable to global competition. But they will have to defend their positions carefully, unlike Harley–Davidson and Norton–Villiers–Triumph, the US and

UK motorcycle makers which suffered so badly at the hands of Honda and Yamaha.

This may seem like trying to have it both ways. It is. As *In Search of Excellence* so elegantly reminded everyone, it is critically important to be able to manage ambiguity and paradox. In Scott Fitzgerald's words, so aptly quoted by Peters and Waterman, 'the test of a first-rate intelligence is the ability to hold two opposed ideas in mind at the same time and still retain the ability to function'.[93]

Having thrown the marketing world, and especially the advertising community, into turmoil with his globalization manifesto, Levitt proceeded to cause consternation among his most fundamentalist followers by admitting that his observations about rampant homogenization contained a good deal of exaggeration.[94] He also conceded that the manifesto had failed to make a quite basic and crucial distinction between products, their brand names and the way they are sold. Even if the same product is sold in different countries, its branding, positioning, promotion and selling need not be identical. They *may* be, as Levitt advocated, but the decision is by no means automatic.

'All I'm really trying to do is to stress the need for companies to examine the growing *similarities* between consumer preferences, as well as the *differences* which still persist', explained Levitt after the manifesto's publication. '*Of course* I'm exaggerating.' If you're trying to change human behaviour, you don't present people with convoluted or judiciously balanced arguments, he maintained. 'When it comes to implementing the ideas in my article, I assume that the reader is someone of commonsense and prudence.'

The trouble was that quite a number of Levitt's readers took him literally. Hence the controversy which exploded in the advertising world after his manifesto's publication, and the near-outraged response of Levitt's friend and fellow marketing guru, Philip Kotler.

Kotler complained that Levitt was 'setting marketing back', by trying to bend consumer demand to suit the product, rather than vice versa.[95] Having spent the previous two-and-a-half decades persuading companies to put market considerations first, and the product second, Levitt was now bent on 'going back to sales'. Kotler was concerned that Levitt's message would rejustify the approach which got international companies into such trouble in the past.

Rather than anticipating the narrowing range of products which Levitt predicted, Kotler maintained that the reverse was occurring. Many new life styles were emerging and new, differentiated markets were opening up. Companies needed a *wide* range of products, and a *wide* range of messages to the consumer, not the reverse.

Kotler's criticism was not softened by Levitt's clarification of his article's rather unclear stance on market segmentation versus homogenization. 'Globalization does not mean the end of market segments', Levitt argued more emphatically than he had in the article itself, but 'that they expand to worldwide proportions'. (He cited, for example, the widespread availability on both sides of the Atlantic of pitta bread, lasagna and Chinese food.)

To which Kotler replied that only a very small proportion of the world's products would be able to be branded globally. One might add that most of the segments which Levitt cited are supplied by regional, national or even local companies, not by global giants, and that their positioning varies from country to country. Most Britons would laugh, for example, at the way that one of their most down-market foods, 'Scotch eggs' (boiled eggs wrapped in sausage meat) have been sold in parts of New England as prime delicacies.

If one allowed for the admitted overstatement of Levitt's article, for his subsequent clarifications and for his jibe that 'Kotler's only looking at the present, not the future', the difference between the two became largely one of degree.

But this still left as a cause for concern the remarkable speed with which a fundamentalist interpretation of Levitt's manifesto had caught the imagination of advertising agencies and their clients – the marketing and product managers of many multi-nationals.

First Saatchi and Saatchi, the then fast-growing British upstart of the international advertising world, set the bandwagon in motion by positioning itself on both sides of the Atlantic as the agency which could help clients seize what it called 'the opportunity for world brands'. Like Levitt's article, it seemed to assume – wrongly – that standardized, global products should automatically be positioned, promoted, advertised and sold in a standardized fashion. Then other, more established international agencies nailed their colours to the same mast – one even claimed that over a fifth of the combined population of France, Britain and West Germany now constituted four new 'global constituencies'.

All this illustrated the dilemma confronting agencies which had handled global campaigns for years, but which had been left floundering by Saatchi's pre-emptive strike. They were presented with the unenviable task of demonstrating their global expertise while still preserving their reputation for sophisticated understanding of the predominant need for market-by-market differentiation in positioning and advertising, if not in the product itself.

The same temptation – of abandoning ambiguity in favour of centralization – and standardization – continues to confront the majority of leading multinational corporations. Ideally, their managements should be trying to find ways to centralize parts of the business system and decentralize others, to 'go global' in some cases and not in others. Doing so is easier said than done. For one thing, it requires them to learn how to strike a series of different organizational balances across each product line and division, between global integration at one extreme and national/regional responsiveness at the other.[96] It also requires each of these organizational balances to be capable of changing from case to case and time to time.

Constructing the sort of flexible 'transnational' which is needed to operate in this way will be one of the hardest management tasks of the 1990s.[97] It is far easier, and more tempting, to throw awkward paradox and diversity to the winds, and to insist on inflexible standardization.

13 Conclusion:
A Two-edged Sword

By rights, the globalization of products and markets should provide an extra stimulus to the emergence of industrial design. In many ways, it strengthens the myriad of influences which, as previous chapters have shown, are prompting a growing number of companies to stop treating designers as mere stylists, and instead to start exploiting their potential to the full.

Sony, Ford, Philips and the other new design 'converts' have gained a distinct competitive advantage by allowing their designers unprecedented involvement in a whole range of decisions about product and market strategy. In principle, globalization should take this process further. The designer's position inside such companies should be enhanced, and the steady stream of new converts should be swollen to a flood.

The trouble is that right does not always triumph, and principles are not always borne out in practice. Existing deterrents against the fully-fledged use of industrial design in many companies could take on new significance if globalization is managed badly. Design would then be pushed back to the dark ages of skin-deep styling, and the companies would be deprived of that 'meaningful distinction' which, as Theodore Levitt rightly argues, is so crucial to the creation of competitive advantage in an era of crowded markets and global competition.[98]

Opportunities

The positive side of the picture is certainly promising. In many companies, the industrial designer remains the only person directly in touch with both technology and the consumer. Despite the introduction of product managers and various other types of coordinator, he or she is often the only person involved with a new product throughout the entire development and production process, from concept to market launch. Equipped with uniquely interdisciplinary attitudes and skills, the designer sits at the centre of a multidimensional matrix, with an eye (and an influence) on every dimension. With the arrival of shorter development cycles, and the need radically to improve communication between different departmental specialists, his or her multidimensional skills become even more invaluable and influential.

These influences, all of them internal to the company for which the designer works, whether in-house or as a consultant, are augmented by a set of changes in the world outside – especially in the marketplace. The slowing of world growth, the intensification of competition, and a host of other factors (examined in chapter 4) have combined to force companies to become more resourceful in distinguishing their products from those of their competitors.

As a result, the creation of real differentiation has become even more important than in the past. But the old weapons for achieving this have become inadequate. No longer can comparative advantage be sustained for long just through lower costs, or higher technologies. Instead, companies are having to emulate the Japanese in adopting sophisticated marketing strategies, in which extremely fine-tuned market segmentation plays a central role.

This process is being accelerated by the onset of globalization. By no means all products, and fewer markets, are becoming globalized. But where they are, the importance of national differences is beginning to decline. Rather than producing an amorphous mass market across the world, as Levitt's fundamentalist followers have wrongly claimed, this process is going hand-in-hand with the splintering of markets into specialist segments, such as young individuals, young families, working women, old people, and so on.[99] To reiterate the vital point made in the previous chapter, globalization does not mean the end of market segments, but their expansion to worldwide proportions. Far from declining, the number of segments may actually increase, partly thanks to social trends and partly because of the efforts of many companies to use further segmentation and differentiation as

'Provided the designer can become computer-literate, he or she has the opportunity to gain still greater influence.'

weapons to defend themselves against the few global mammoths which are hell-bent on worldwide homogenization.

This is no simple matter of designing standard products, and selling them to different segments in different ways. Instead, it requires the introduction of marketing proper: that is, the sensitive understanding of various customer groupings, and the design of different products to suit their various preferences. Hence the need for many companies to convert to the full use of marketing and design in tandem. Though industrial designers frequently can – and do – substitute for the absence of marketing imagination, in most companies the most potent force for imaginative marketing and product strategy is a real partnership between marketing and design. It is just such a partnership which has given vital extra strength to the global strategies of Japanese innovators such as Olympus, Sony and, more recently, Honda. The example set by Japan's design-minded companies is one of the many factors which are galvanizing Western enterprises into their own belated upgrading of design.

A further set of opportunities for industrial design is created by emerging trends in the technology of both products and processes.

In many product areas, especially consumer electronics, technology has become so miniaturized and standardized that it amounts to little more than a few microchips – what the electronics industry calls a 'black box'. As a result, the product's function no longer has much, if any, influence on its form. The technological heart of the product has become a module which allows the designer almost complete freedom to design the product as he or she sees fit – or, more appropriately and less arrogantly, as the target customer would most prefer it. In such circumstances, the possibilities of different forms and functions, and combinations between them, have become now almost endless.

At the same time, the saturation of certain markets (such as radio and TV), and the convergence of different technologies (such as audio, video and computer communications) is creating the opportunity – and the demand – for more complex 'system products'. Initially this involved relatively simple combinations, such as radio-recorders and the use of more hi-fi audio technology in video products. But the trend has since moved a complex stage further, to complete home entertainment systems. Instead of stand-alone radios, TVs, video recorders and home computer terminals, the individual units are being combined, or developed as modules which can be gradually pieced together, as the consumer's preference (and pocket) allow. In a company such as Sony or Philips this requires an unprecedented degree of collaboration between functional specialists, and between different divisions. Never have the synthesizing and communication skills of the industrial designer been more invaluable. Hence, for example, the promotion of Sony's design chief in 1985 to become 'Director of Consumer System Products and Design', and then in 1988 to join Sony's main executive board.

A revolution is also affecting the process technologies of design, development and production. The ramifications of computer-aided design and manufacture (CADCAM) for the management of product development are so far-reaching that a separate study would be required to do them justice. The same applies to the impact of flexible manufacturing systems (FMSs) on development, production and competitive strategy. Suffice to say that together they offer the opportunity to: design and develop products more rapidly than in the past; tailor designs to target market segments much more precisely and economically; radically improve the process of design-for-manufacture, so that expensive and time-consuming rework can be avoided; and lower the unit cost of short production runs.

For industrial design the implications obviously include the opportunity to create more varied products. But they are also more fundamental. The use of common computer data to originate designs and manufacturing processes requires the company to lower the barriers between different specialist functions, so that industrial designers, product engineers and manufacturing engineers can work together much more closely than before. Provided the designer can become computer-literate, as well as more conversant with engineering and production, he or she has the opportunity to gain still greater influence over the product development process.

Threats

Against this highly positive outlook for industrial design must be set a number of contrary pressures, some of which arise from precisely the same trends which offer more scope for design. The outcome depends partly on design's ability to exert its muscle, but primarily on the ability of top management to see where the company's real interest lies.

Prime among these double-edged swords is the combined effect of shorter development cycles, of CADCAM, and of the growth of 'black box' electronics. Just as the lowering of departmental barriers associated with shorter cycles and CADCAM offers the opportunity for industrial design to gain greater influence over various breeds of engineer, so the same applies in reverse. Along with the blurring of frontiers between departments will inevitably come a shift in the balance of power between the different functions. As has been argued in previous chapters, industrial design has all the credentials to claim that its position should be reinforced, and it is in the broader interest of the company to make sure this occurs. But it is not a foregone conclusion. Especially in companies which have been dominated by scientist–engineers, or by a production mentality, there is a risk that other interests may prevail.

'Black box' technology clearly offers unrivalled scope to combine different forms and functions into a wide range of products, and thereby to satisfy the preferences of a broad range of different market segments. But it also provides the temptation to make everything the same, or at least to confine product differentiation just to shapes and colours. When Philips or one of the Japanese electronics majors uses a standard set of circuit boards and tape mechanisms in several ranges of hi-fi products targeted at very different market segments in Europe, the US and Japan, it becomes all too easy to fall into the trap of homogenizing the product to an excessive degree, thereby doing a disservice both to the company's customers and to the potential of design. Until flexible manufacturing systems are in widespread use, the pressure for massive economies of scale in component production and final assembly will continue to tempt companies to over-standardize their products.

Several other factors also increase the danger of homogenization. Especially in office products, the growth of international standards for colour, shape, and various other aspects of ergonomics is leaving less and less room for differentiation. The stranglehold of powerful, often conservative, retailers is another constraining influence, as is the aversion of many manufacturers to innovation. Here, as outlined in chapter 3, the problem is often rooted in myopic sales thinking, masquerading as marketing. It is a sad comment on the degree of copying which now pervades the world of consumer electronics that the industry's prime innovator had to start differentiating some of its products from those of its emulators by decorating them with a label saying 'It's a Sony'.

These pressures risk being increased by globalization. If, regardless of the interests of their customers, companies respond to the global challenge with crass and insensitive strategies of standardization right through every link in their value-added chain, the risk will become reality. It must be hoped that most companies will make a careful, link-by-link assessment of the potential benefits and disadvantages of globalization, and then adopt different strategies for each link: research; design; development; component supply and manufacture; product assembly; marketing; branding; sales and distribution; and so forth. If the majority of companies do adopt this approach, globalization will be confined within sensible limits, and will indeed prove an opportunity rather than a threat, both for the consumer and for the industrial designer.

Prospects

On balance, the prospect must be that common sense will prevail, and that the number of corporate design converts will, therefore, continue to grow apace. Some will focus all their initial attention on better product design, others will follow Olivetti and IBM in rapidly extending their design commitment to visual communications, factories, shops and offices. Many will use design consultants as their main source of expertise, while other companies will rely mainly on in-house staff. A number will use industrial design as a partial substitute for marketing, while many others will forge a new partnership between the two, as this book advocates.

As was pointed out in part II ('Design Unchained'), the choice of route depends not only on the company's particular situation, but also on its culture. The new corporate design converts share few characteristics, other than a basic trio of common attributes: the need to upgrade design, and give it an independent voice; the pivotal role played by a few key personalities in creating the necessary change; and the importance of top management commitment.

As Sony, Ford, Philips and all the other converts have discovered, the creation of a strong design dimension is an important part of the battle to satisfy the customer and beat the competition. But it does not develop of its own accord. Like everything else in this world, it has to be designed.

Appendix:

How Japan 'Cascades' Through Western Markets

The competitive strategy of Japanese companies is much more consistent and systematic than many threatened Western companies realize.* Take the example of consumer electronics. Japan's invasion has not been achieved in a series of unrelated market segments, but through a complementary series of steps, from the transistor radio of 30 years ago to today's videotape player.

The only effective way for Western companies to defend themselves against future attack is to borrow or trump the Japanese approach. Marc Particelli, a vice-president of Booz, Allen & Hamilton, the management consultancy, warns 'your industry may be next to come under attack'.

Tracing Japan's gradual penetration and dominance of the US consumer electronics market, Particelli describes what he calls a 'cascading pattern', beginning with carefully selected small segments and gradually moving across the entire market. This applies both within broad product markets (transistor radios, for example) and within consumer electronics as a whole.

Examining this process in detail, Particelli concludes that a number of critical success factors underpin Japanese strategy. They include:

(1) Initial penetration of well-defined target segments

As a rule, the Japanese start with a large business in their highly protected home market, enter peripheral markets – in

*This is a summary of an article by Marc Particelli which was published in Booz Allen's journal *Outlook* (No. 4, 1981). It was first reported in *The Financial Times* on 9 November 1981.

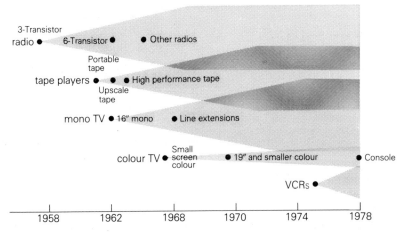

Figure 3 Japan's expansion strategy: by a 'cascading' penetration of related market segments (adapted from Booz, Allen & Hamilton diagram).

the advanced developing countries, for instance – and then take dead aim at US and European markets. Their initial US penetration is always at an extremely well-defined target segment, with a limited line – typically at the low end of the product category. They pick off volume channels where economical distribution can be achieved, with little concern for whether they market private label or branded products. Usually they rely on retail push and private labels rather than on investments in marketing for brand 'pull'.

(2) Volume stimulation and segment domination

Their initial base secured, the Japanese then proceed to stimulate volume and dominate that particular segment. This achieved, they move on to another segment and repeat the process, generally concentrating on providing consumers with low price, and always with extremely high quality, driving continually for price and cost (reduction) to increase consumer support and stimulate growth.

(3) Resource application: product and cost vs. innovation

The Japanese support their emphasis on value in price, benefits and quality by applying resources in a way that differs dramatically from their US and European competitors.

In general, the early emphasis in resource allocation by US manufacturers is on innovation: the invention of new products and applications, and product improvements. As a market begins to mature, emphasis is placed on market stimulation through marketing and sales. Only at the mature stage does the American manufacturer begin to focus a disproportionate share of his or her resources on cost (reduction).

(4)   Establishing consumer value and building market presence

As a key element in their strategy, the Japanese have consistently relied on private brands, letting them assume marketing risks and then building their own brand presence from this secure private label base. By taking this approach they have established a strong consumer franchise; most buyers worldwide now consider Japanese products to be superior in quality and value to US and European ones.

(5)   Cost cutting all fronts

In order to achieve both low retail price and superior retailer margins, the Japanese need to deliver their product at a low wholesale price. So they put strong emphasis on economies in all major cost areas: manufacturing, distribution, marketing. This emphasis on all costs, from production through consumer purchase, is another key element in Japanese strategy that distinguishes it from that of American and European firms.

(6)   Implementation in a global marketplace

Japan has developed the ability to look at the entire world as a potential marketplace. Starting from the home base, it moves on to a global level in order to realize sufficient economies of scale. The Japanese tendency to enter Western markets in a concentrated geographic area, and in a segment which is small and frequently unimportant to domestic companies, provides the basic knowledge, people, systems and customers from which to build broad acceptance and capture market share. These early steps should be a warning signal for domestic companies.

References

1 Olins, W., *The Corporate Personality*, London, 1978.
2 Olins, W., 'Management by Design', *Management Today*, London, February 1985.
3 Olins, W., *Corporate Identity*, London, 1989.
4 Papanek, Victor, *Design for the Real World*, New York, 1971 and London, 1972. Also (Papanek and various authors) *Design and Society*, Volume One of Proceedings of Design Policy Conference at the Royal College of Art, 1982.
5 Kotler, Philip and Roth, G. Alexander, 'Design: A Powerful Strategic Tool', *Journal of Business Strategy* (US), Autumn 1984.
6 *Ibid.*
7 Clark, Kim and Fujimoto, Takahiro, 'Reducing the Time to Market: The Case of the World Auto Industry', *Design Management Journal*, Boston, Vol. 1, No. 1, 1989. Also, Fujimoto, Takahiro, 'Product Integrity and the Role of Industrial Design', Unpublished discussion paper, Harvard Business School, October 1989.
8 Levitt, Theodore, *The Marketing Imagination*, New York and London, 1983, p. 128.
9 Clark and Fujimoto, 'Reducing the Time to Market'.
10 Pulos, Arthur J., *The American Design Ethic*, Boston, 1983, p. 260.
11 Heskett, John, *Industrial Design*, London, 1980, pp. 118–19.
12 *Ibid*, pp. 120–6.
13 Sparke, Penny, *Consultant Design*, London, 1983, p. 24.
14 'Both Fish and Fowl', *Fortune*, New York, January 1934.
15 Pulos, *American Design Ethic*, p. 279.
16 Bayley, Stephen, *In Good Shape*, London, 1979, p. 19.
17 Bayley, Stephen, *Harley Earl and the Dream Machine*, London, 1983.
18 Pulos, *American Design Ethic*, p. 289.
19 Dreyfuss, Henry, *Designing for People*, New York, 1955, p. 69.

20 *Ibid*, p. 77.
21 Sparke. *Consultant Design*, pp. 35–8.
22 Gorb, Peter (ed.) *Living by Design*, London, 1978, p. 255.
23 Levitt, *The Marketing Imagination*, p. 130.
24 Galton, Francis, *Inquiries into Human Faculty*, London, 1883, 2nd edition, 1919.
25 Dreyfuss, *Designing for People*, p. 57.
26 'Why Italian Industrial Design is Sweeping the World', *Business Week*, New York, 3 September 1984.
27 Caplan, Ralph, *By Design*, New York, 1982.
28 Levitt, Theodore, 'Marketing Myopia', *Harvard Business Review*, 1960. Republished in *Product Policy for Consumer Goods Companies*, HBR, Boston.
29 *Ibid*, p. 14.
30 Kotler, Philip, Fahey, Liam, and Jatusripitak, S., *The New Competition*, New Jersey, 1985, ch. 3.
31 'Marketing: the New Priority', *Business Week*, 21 November 1983. Also 'Hewlett-Packard Discovers Marketing', *Fortune*, 1 October 1984.
32 Peters, Thomas and Waterman, Robert, *In Search of Excellence*, New York, 1982, p. 156.
33 Doyle, Peter *et al. A Comparative Investigation of Japanese Marketing Strategies in the British Market*, Bradford Management Centre, 1985.
34 Unpublished research by leading European business academics and management consultants.
35 Kotler, Philip, *Creating a Marketing Culture*, Stockton Lecture at London Business School, March 1985. Published in *LBS Journal*, **10**, 1.
36 Levitt, 'Marketing Myopia'.
37 *Ibid*, p. 10.
38 *Ibid*.
39 Kotler, *The New Competition*, pp. 43–58.
40 *Ibid*.
41 *Ibid*.
42 *Ibid*.
43 Levitt, 'Marketing Myopia', p. 14.
44 Kotler, Philip, 'From Sales Obsession to Marketing Effectiveness', *Harvard Business Review*, November–December 1977.
45 King, Stephen, 'Applying Research to Decision-Making', *LBS Journal*, London, Winter 1983, p. 6.
46 Rosenbloom, Richard and Abernathy, William, 'The Climate for Innovation in Industry', *Research Policy* (UK), **II**, 4. 1982–3.
47 Levitt, 'Marketing Myopia', p. 15.

48 Day, George S. and Wensley, Robin, 'Marketing Theory with a Strategic Orientation', *Journal of Marketing (US)*, **47** (Fall 1983), pp. 79–89.
49 Simmonds, Kenneth, 'Peaks and Pitfalls of Competitive Marketing', Stockton Lecture at London Business School, January 1985. Published in *LBS Journal*, **10**, 1.
50 Rosenbloom and Abernathy, 'Climate for Innovation'.
51 *Ibid.*
52 Bonoma, Thomas, 'Making your Marketing Strategy Work', *Harvard Business Review*, March–April 1984.
53 Simmonds, Kenneth, 'Peaks and Pitfalls of Competitive Marketing'.
54 *Fortune*, 1 October 1984.
55 Jobs, Stephen, Paper to International Design Conference, Aspen, 1983.
56 Hayes, Robert, Clark, Kim and Lorenz, Christopher (eds), *The Uneasy Alliance: Managing the Productivity–Technology Dilemma*, Boston, 1985, Part III.
57 *Ibid*, pp. 337–81.
58 Sasser, Earl and Wasserman, Dale, unpublished Harvard Business School Working Paper.
59 Olivetti Direzione di Corporate Image, *Olivetti Design Process*, 1908–83, p. 2.
60 'Das Geheimnis der roten Kassetten', *Manager Magazin*, 12/82, Hamburg.
61 Viti, Dr. Paolo, comments at Design Management Institute Conference, London, May 1982.
62 Broehl, Wayne G. Jr., *John Deere's Company*, New York, 1984, p. 531.
63 *John Deere Tractors 1918–76*. Moline, p. 10.
64 Maxon, Ira, Letters to Henry Dreyfuss, 22 December 1938. Deere & Co. Library.
65 Broehl, *John Deere's Company*, pp. 639–40.
66 *Ibid.*
67 Purcell, William, 'Industrial Design – a Vital Ingredient', *Automotive Industries (US)*, 15 May 1961.
68 Interview with Dr Gordon Millar, Deere & Co., Moline, May 1984.
69 Morita, Akio, Speech at Sony Design Exhibition, Boilerhouse Project, Victoria and Albert Museum, London, Spring 1982.
70 Described in exhibition catalogue, *Sony Design*, Boilerhouse Project, London.
71 Interview with Yasuo Kuroki, New York, June 1983.
72 Interview with Yasuo Kuroki, Boston, May 1984.
73 Peters, Tom, *Thriving on Chaos*, New York, 1987.

74 Lutz, Robert, 'Ford pins its hopes on Sierra', *The Financial Times*, 21 September 1982.
75 *Ibid.*
76 Interview with Uwe Bahnsen, Dunton, November 1982.
77 'Philips in Search of Speed', *The Financial Times*, 25 February 1985.
78 Interviews with Mike Jankowski, Eindhoven, August 1983 and January 1985.
79 Interview with Cor van der Klugt, Eindhoven, August 1983.
80 See Heskett, John, *Philips: a Study of the Corporate Management of Design*, London, 1989.
81 Smith, M. R. H., paper to National Conference on Quality and Competitiveness, London, November 1981. Reported in *The Financial Times*, 25 November 1981.
82 Interview with Charles McCaskie, Peterborough, October 1983.
83 Woudhuysen, J. (ed.) *Central to Design – Central to Industry*, London 1982.
84 'Kenneth Grange at the Boilerhouse', May 1983, Exhibition Catalogue, p. 25.
85 Grange Exhibition Catalogue, pp. 36–40.
86 Grange Exhibition Catalogue, pp. 12, 29.
87 Interview with Geoffrey Place, Cincinnati, June 1984.
88 Ohmae, Kenichi, *Triad Power: the Coming Shape of Global Competition*, New York and London, 1985.
89 Levitt, Theodore, 'The Globalization of Markets', *Harvard Business Review*, May/June 1983. Also longer version in *The Marketing Imagination*, New York, 1983.
90 Porter, Michael, *Competitive Strategy*, New York, 1980 and *Competitive Advantage*, New York, 1985.
91 Porter, Michael, 'Competition in Global Industries: a Conceptual Framework', *Briefing Book for Harvard Business School Colloquium on Competition in Global Industries*, April 1984, p. 27.
92 Quelch, John, Presentation to BBDO Seminar on Globalization, London, June 1984.
93 Peters and Waterman, *In Search of Excellence*, p. 89.
94 Interview with Theodore Levitt, Boston, July 1984.
95 Interview with Philip Kotler, Chicago, July 1984.
96 Doz, Yves and Prahalad, C.-K., *The Multinational Mission*, New York and London, 1987.
97 Bartlett, Christopher A. and Ghoshal, Sumantra, *Managing Across Borders*, Boston and London, 1989.
98 Levitt, *The Marketing Imagination*, p. 128.
99 Blaich, Robert, 'Design Management in a Global Corporation', *Innovation* (Journal of the Industrial Designers Society of America), McLean, VA, Spring 1985.

Illustrations Acknowledgements

The author and publishers would like to thank the individuals and organizations mentioned below who kindly supplied the illustrations which appear on the preceding pages, and with whose permission they are reproduced:

Booz, Allen & Hamilton: p. 159
Deere & Co: pp. 70, 74, 77
Henry Dreyfuss Associates: p. 19
The Financial Times: p. 133 (FT photo by Trevor Humphries); p. 143 (FT photo by Alan Harper)
Ford Motor Company: pp. 5, 6, 7, 97, 99, 102, 104
McKinsey & Co: p. 40
Olivetti: pp. 57, 59, 61, 65
Pentagram: pp. 128, 129, 130, 131, 132, 135
Philips: pp. 25, 112, 115, 116
Sony: pp. 83, 85, 87, 90
John Springs: p. 28

Particular thanks are due to Mike Daley for his characteristically fine drawings, specially commissioned for the book; also to Denis Kiley of *The Financial Times* for arranging release of FT copyright on material which forms the basis of several of the chapters.

Index

Index by Meg Davies